AN INTRODUCTION TO CUT-OFF GRADE ESTIMATION

BY JEAN-MICHEL RENDU

Published by

Society for Mining,
Metallurgy, and Exploration, Inc.

Society for Mining, Metallurgy, and Exploration, Inc. (SME)
8307 Shaffer Parkway
Littleton, Colorado, USA 80127
(303) 973-9550 / (800) 763-3132
www.smenet.org

SME advances the worldwide mining and minerals community through information exchange and professional development. SME is the world's largest association of mining and minerals professionals.

ISBN 978-0-87335-268-0

Library of Congress Cataloging-in-Publication Data
Rendu, J. -M.
An introduction to cut-off grade estimation / by Jean-Michel (J.M.) Rendu. -- 1st ed.
p. cm.
Includes bibliographical references and index.
ISBN 978-0-87335-268-0
1. Ores--Grading. 2. Ores--Sampling and estimation. I. Title.

TN560.R45 2008
622'.7--dc22

2008036206

Contents

	PREFACE	v
CHAPTER 1	INTRODUCTION	1
CHAPTER 2	GENERAL PRINCIPLES	5
	Mathematical Formulation	5
	Cut-off Grade and Grade–Tonnage Relationship	6
	Direct Profit and Loss	7
	Opportunity Costs and Benefits	9
	Cut-off Grade Optimization with Opportunity Costs	14
	Other Costs and Benefits	15
CHAPTER 3	MINIMUM CUT-OFF GRADES	19
	Cut-off Grade Between Ore and Waste	19
	Cut-off Grade for Material at the Bottom of an Open Pit Mine	22
	Cut-off Grades in Underground Mines	24
	Cut-off Grade to Choose Between Processes	26
	Cut-off Grade Between Waste and Low-grade Stockpile	28
	Cut-off Grade with Variable Recoveries	30
	Opportunity Cost of Not Using the Optimum Cut-off Grade	33
CHAPTER 4	CUT-OFF GRADE FOR POLYMETALLIC DEPOSITS	37
	General Considerations	37
	Calculation of Cut-off Grades Using Net Smelter Return	38
	Calculation and Reporting of Metal Equivalent	40
CHAPTER 5	CUT-OFF GRADE AND OPTIMIZATION OF PROCESSING PLANT OPERATING CONDITIONS	43
	Mathematical Formulation	43
	Example: Optimization of Grinding Circuit in a Copper Mine	45
CHAPTER 6	CUT-OFF GRADE AND MINE PLANNING—OPEN PIT AND UNDERGROUND SELECTIVE MINING	53
	Open Pit Mine: Economic Valuation of a Pushback	53
	Underground Mine: Economic Valuation of a Stope	54
	Similarities Between Open Pit and Underground Mine Planning	55

CHAPTER 7 CUT-OFF GRADE AND MINE PLANNING—BLOCK AND PANEL CAVING 57
Constraints Imposed by Block and Panel Caving 57
Marginal Cut-off Grade and Draw Point Management 58
Marginal Cut-off Grade and Block Design 58
Influence of Capital Cost and Discount Rate 59
Opportunity Cost 60

CHAPTER 8 WHICH COSTS SHOULD BE INCLUDED IN CUT-OFF GRADE CALCULATIONS? 63

CHAPTER 9 WHEN MARGINAL ANALYSIS NO LONGER APPLIES: A GOLD LEACHING OPERATION 67

CHAPTER 10 MINING CAPACITY AND CUT-OFF GRADE WHEN PROCESSING CAPACITY IS FIXED 71

CHAPTER 11 PROCESSING CAPACITY AND CUT-OFF GRADE WHEN MINING CAPACITY IS FIXED 75

CHAPTER 12 MINING AND PROCESSING CAPACITY AND CUT-OFF GRADE WHEN SALES VOLUME IS FIXED . . 79
Fixed Sales with No Mining or Processing Constraint 79
Fixed Sales and Fixed Processing Rate with No Mining Constraint 80
Fixed Sales and Fixed Mining Rate with No Processing Constraint 82

CHAPTER 13 RELEASING CAPACITY CONSTRAINTS: A BASE METAL EXAMPLE 85

CHAPTER 14 RELATIONSHIP BETWEEN MINE SELECTIVITY, DEPOSIT MODELING, ORE CONTROL, AND CUT-OFF GRADE 89

CHAPTER 15 CONCLUSIONS 93

BIBLIOGRAPHY 95

SYMBOLS 97

INDEX 101

ABOUT THE AUTHOR 105

Preface

This book started with a desire to understand how to answer an apparently simple but actually complex question faced by all those responsible for the development and operation of mines: How do we determine which cut-off grade should be used to separate material that should be processed from that which should be sent to the waste dump? The answer appears straightforward: If it is profitable to process one metric ton of material, this ton should be processed. But what is *profitable*? The cut-off grade has a direct bearing on the tonnage of material mined, the tonnage and average grade of material processed, the size of the mining operation, and consequently capital costs, operating costs, and environmental and socio-economic impacts. Should we maximize cash flow, net present value, the life of the mining operation, the return to shareholders? How do we take into account economic, environmental, social, political, ethical and moral values, objectives, and regulations?

Somewhat surprisingly, only one other book has been written exclusively on the subject of cut-off grade estimation: *The Economic Definition of Ore: Cut-Off Grades in Theory and Practice* by Ken Lane, published in 1988. Lane's book was and will remain the standard for mathematical formulation of solutions to cut-off grade estimation when the objective is to maximize net present value. Concepts first formulated by Lane were used as the foundation of this book.

Considerable progress has been made in the last twenty years to improve mine planning and optimize cut-off grades. Increasingly complex algorithms have been developed, and better, easier to use computer programs have been written to assist engineers and economists in analyzing mine plans, testing the options, and improving production schedules. Computer programs have become easier to use, but the assumptions made by those who write the programs are often lost to the end user. With this book I am hoping to bridge the gap between theory and practice, the ivory tower and engineers in the field, by describing the fundamental principles of cut-off grade estimation and providing concrete examples.

This book started as notes written during the last thirty years. Eventually these notes turned into an introductory short course. Each time I gave the course, more and more questions were asked concerning increasingly complex situations, demanding more practical examples and challenging the assumptions made. Each question resulted in corrections, additions, and more chapters. I am extremely thankful to those who helped me in this respect. They include too many individuals over too many years to be listed here. They know who they are and I would not have continued this work without their probing and their interest in the subject. I am particularly grateful to Ernie Bohnet who kept on motivating me when I doubted that I had a story to tell or that there would be sufficient interest in continuing this effort to make it worthwhile. It is because of Ernie that I completed this book. I also want to thank the Society for Mining, Metallurgy, and Exploration, Inc., and Jane Olivier, who accepted the manuscript and brought it to publication in record time. None of these people, of course, can be blamed for any errors or lapses that I may have made and for which I am fully responsible.

My first book, *An Introduction to Geostatistical Methods of Mineral Evaluation*, was published in 1978 with the objective to clarify the already arcane science of geostatistics. It is only fitting that *An Introduction to Cut-Off Grade Estimation* be published, with similar objectives, in 2008, exactly thirty years later.

DEDICATION

I am dedicating this book to my wife Karla and my children, Yannick and Mikael. Life with a husband and father who spent too much time traveling to remote mines all over the world, and then returned home to work long hours in front of his computer, was not without challenges and disappointments. I am grateful for their patience, understanding, and unquestioning love.

CHAPTER ONE

Introduction

A *cut-off* grade is generally defined as a minimum amount of valuable product or metal that one metric ton (that is, 1,000 kilograms) of material must contain before this material is sent to the processing plant. This definition is used to distinguish material that should not be mined or should be wasted from that which should be processed. Cut-off grades are also used to decide the routing of mined material when two or more processes are available, such as heap leaching and milling. Cut-off grades are used to decide whether material should be stockpiled for future processing or processed immediately.

Cut-off grades are calculated by comparing costs and benefits. In simple geological and metallurgical environments, a single number, such as a minimum metal content, is sufficient to define the cut-off grade. In most situations, costs and recoveries, and therefore cut-off grades, vary with the geological characteristics of the material being mined. Grade is usually the most important factor but may not be the only one. If material is sent to a waste dump, the acid-generating potential of this material may have a direct impact on costs related to environmental controls. Sulfide content may be a critical—even overriding—factor for material sent to a roasting or flotation plant. Clay content may have a deleterious effect on the recovery and throughput of a leaching plant.

The cut-off grade defines the profitability of a mining operation as well as the mine life. A high cut-off grade can be used to increase short-term profitability and the net present value of a project, thereby possibly enhancing the benefit to shareholders and other financial stakeholders, including government and local communities. However, increasing the cut-off grade is also likely to decrease the life of the mine. A shorter mine life can reduce time-dependent opportunities, such as those offered by price cycles. A shorter mine life can also result in higher socio-economic impact with reduced long-term employment and decreased benefits to employees and local communities.

Increased cut-off grades may be considered to reduce political risk by ensuring a higher financial return over a shorter time period. The cut-off grade may be increased when metal prices increase if this is needed to strengthen the financial position of the company and reduce the risk of failure when metal prices fall. Conversely, cut-off grades may be decreased during periods of high prices to increase mine life and keep high-grade material available to maintain profitability when prices fall. Cut-off grades may also be constrained by economic or technical performance criteria imposed by banks and other financial institutions.

In some instances, a conscious decision might be made to increase the mining capacity while keeping the processing capacity constant. This allows an increase in cut-off grade. Some of the lower-grade material may be stockpiled for processing at a later date. Stockpiling may have a number of consequences—some positive (such as increased useful life of processing facilities) and others negative (such as increased environmental risk and decreased metallurgical recovery of stockpiled material).

Cut-off grades have a direct impact on reserves for which the public release is subject to the rules and regulations of the various stock exchanges and other regulatory agencies. Published reserves and generally accepted accounting practices are linked. Reserves enter into the calculation of capital depreciation, company book value, unit cost of production, and taxes. Published reserves are also linked to the value that the financial market gives to a mining company. For some commodities, there is a fairly widely held but arguably incorrect belief that this link is primarily a function of the magnitude of the reserves and that quality is of lesser significance. Low cut-off grades may be considered desirable by those calculating or publicly reporting reserves if personal bonuses are a function of the magnitude of the published reserves. As a result of these various links—some desirable, some not—it may seem desirable to maximize the published reserves by using the lowest technically, financially, and legally defendable cut-off grade. However, one must always keep in mind that reserves are published to inform investors and other stakeholders, and that processes and controls should be put in place to eliminate the influence of factors that could result in publication of misleading estimates.

Both outsider and insider stakeholders have an interest in the cut-off grade and the reserves deriving from it. Outsiders include shareholders, financial institutions, local communities, environmentalists, regulators, governmental and non-governmental agencies, suppliers, contractors, and buyers of the product being sold. Insiders include company management and employees. The board of directors represents the interests of the shareholders and is often composed of both insiders and independent outsiders. Cut-off grades are and should be calculated primarily by taking into account only technical and economic constraints. However, the often-conflicting interests and objectives of the many stakeholders must be understood and prioritized in order to make the best decision concerning cut-off grade determination.

The technical literature includes many publications on estimating and optimizing cut-off grades. The most comprehensive reference is Kenneth F. Lane's *The Economic Definition of Ore: Cut-Off Grades in Theory and Practice* (refer to the bibliography for publication information). The objective most commonly accepted in cut-off grade optimization studies is to optimize the net present value of future cash flows. To reach this objective, one must take into account space-related variables (such as the geographic location of the deposit and its geological characteristics), as well as time-related variables (including the order in which the material will be mined and processed), and the resulting cash flow. The time–space nature of the problem is quite complex; consequently, so are the proposed mathematical solutions to cut-off grade optimization. The bibliography provides detailed references to some of these solutions. This book attempts to explain basic concepts in a simple fashion, making them accessible to mine managers, analysts, geologists, mining engineers, and other practitioners.

CHAPTER TWO

General Principles

Choosing a cut-off grade is equivalent to choosing the value of a geologically defined parameter or set of parameters that will be used to decide whether one metric ton of material should be sent to one process or another.

MATHEMATICAL FORMULATION

Let x be the value of the parameter(s) that must be taken into account to determine the destination to which the material should be sent. In simple cases, a single parameter may be sufficient to define the destination, such as copper grade or gold grade. In other cases, a set of parameters may have to be considered such as copper and gold grades, sulfide content, clay content, and percentage of deleterious elements.

The value, or *utility*[1], of sending one metric ton of material with parameter value (grade) x to destination 1 (process 1) is $U_1(x)$. The utility of sending the same material to destination 2 (process 2) is $U_2(x)$. The cut-off grade x_c is the value of x for which

$$U_1(x_c) = U_2(x_c)$$

If $U_1(x)$ exceeds $U_2(x)$ for x greater than x_c, then all material for which x is greater than x_c should be sent to process 1.

As indicated in the introduction, the choice of a cut-off grade is governed primarily by financial objectives. However, the consequences of choosing a given cut-off grade are complex and not all of a financial nature. When estimating cut-off grades, all controlling variables must be taken into account. To facilitate this process the utility U(x) of sending material of grade x to a given process is expressed as the sum of three parts:

1 The term *utility* is used in decision theory to represent the satisfaction gained from following a given course of action. This satisfaction is a function of preferences and values specific to the decision-maker. The utility of a given cut-off grade strategy is a measure of the extent to which this strategy reaches the mining company's objectives.

$$U(x) = U_{dir}(x) + U_{opp}(x) + U_{oth}(x)$$

In this equation, $U_{dir}(x)$ represents the direct profit or loss that will be incurred from processing one metric ton of material of grade x. $U_{opp}(x)$ represents the opportunity cost or benefit of changing the processing schedule by adding one metric ton of grade x to the material flow. This opportunity cost is incurred only when there are constraints that limit how many metric tons can be processed at a given time. Other factors that must be taken into account in the calculation of cut-off grades but may not be quantifiable are represented by $U_{oth}(x)$.

CUT-OFF GRADE AND GRADE–TONNAGE RELATIONSHIP

The cut-off grade determines the tonnage and average grade of material delivered to a given process and therefore the amount of product sold. In first approximation, if T_{+c} represents the tonnage and x_{+c} the average grade of material above the cut-off grade x_c, the revenue from sales is equal to $T_{+c} \cdot x_{+c} \cdot r \cdot V$, where r is the proportion of valuable product recovered during processing and V is the market value of the product sold. The cut-off grade also determines the tonnage of material mined that will not be processed. Figure 2-1 shows the relationship between cut-off grade and tonnage and average grade above cut-off grade. The curves on this graph are known as the grade–tonnage curves.

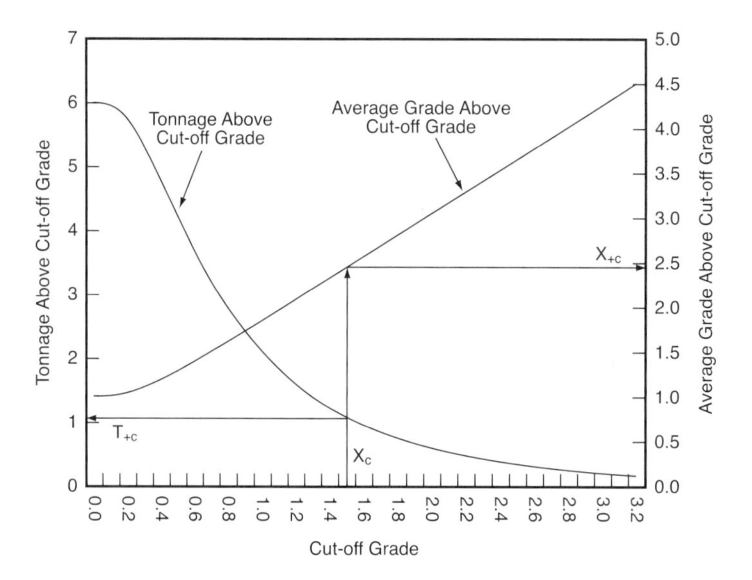

FIGURE 2-1 Example of grade–tonnage curve

Grade–tonnage curves are used extensively throughout this book to illustrate the impact of different cut-off grade strategies on the economics of a mining operation.

DIRECT PROFIT AND LOSS

Direct profits or losses associated with one metric ton of material, $U_{dir}(x)$, are estimated by taking into account only costs and revenues that can be directly assigned to mining this material, processing it, and selling the final product.

Mathematical Formulation

The direct profit or loss $U_{dir}(x)$ expected from processing one metric ton of material of grade x is $U_{ore}(x)$, expressed as follows:

$$U_{ore}(x) = x \cdot r \cdot (V - R) - (M_o + P_o + O_o)$$

- x = average grade
- r = recovery, or proportion of valuable product recovered from the mined material
- V = value of one unit of valuable product
- R = refining, transportation, and other costs incurred per unit of valuable product
- M_o = mining cost per metric ton processed
- P_o = proccessing cost per metric ton processed
- O_o = overhead cost per metric ton processed

If the valuable product is a concentrate, V is the value of one unit of metal contained in the concentrate. For example, V can be the copper price expressed in dollars per pound of copper or the gold price expressed in dollars per troy ounce of gold. The variable r is the percentage of metal in one metric ton of material of grade x that will be recovered and paid for by the buyer. R includes transportation and refining costs, and other deductions and penalties to be deducted from V. When concentrate is sold to a smelter, the applicable values of V and R may be negotiated between seller and buyer and specified in a smelter contract.

If the material is to be wasted, the value of $U_{dir}(x)$ is $U_{waste}(x)$, expressed as follows:

$$U_{waste}(x) = -(M_w + P_w + O_w)$$

M_w and O_w are mining and overhead costs per metric ton of waste. P_w is the cost of processing one metric ton of waste as necessary to avoid potential water contamination and acid generation, and to satisfy other applicable regulatory and environmental requirements. The cut-off grade between ore and waste is x_c, such that $U_{ore}(x_c) = U_{waste}(x_c)$.

Precious Metal Example

To illustrate how these formulae are used to calculate the cut-off grade, consider a gold mining operation with these characteristics:

- For ore being processed, r = 80%, V = $270.00 per ounce of gold, R = $5.00 per ounce, M_o = $1.00 per metric ton mined and processed, P_o = $15.00 per metric ton processed, and O_o = 20% of operating costs.
- For wasted material, $M_w + P_w$ = $1.10, and O_w = 20% of operating costs.

If only direct costs and revenues are taken into account, the cut-off grade between ore and waste is x_c such that the utility of processing one metric ton of material of grade x_c is equal to the utility of wasting this metric ton:

$$x_c \cdot 0.80 \cdot (270.00 - 5.00) - 1.20 \cdot (1.00 + 15.00) = -1.20 \cdot 1.10$$

$$x_c = [1.20 \cdot (1.00 + 15.00) - 1.20 \cdot 1.10]/[0.80 \cdot (270.00 - 5.00)]$$

$$x_c = 0.084 \text{ ounces/metric ton} = 2.62 \text{ grams/metric ton}$$

Base Metal Example

As another example, consider an open pit copper mine. The last pushback is being mined and it's necessary to decide whether material located at the bottom of the pit should be mined and processed or wasted and left in place. The operation is characterized as follows:

- The mining cost is $1.00 per metric ton of ore. The mill processing cost is $3.00 per metric ton processed. Concentrate is produced. Shipping, smelting, and refining costs are $0.30 per pound of fine copper produced.
- The mill recovery is 89% and the smelting recovery is 96.5% for a total recovery of 85.9%.
- The copper price is $1.00 per pound of copper. There are 2,205 pounds of copper per metric ton.
- There is no cost associated with leaving material at the bottom of the pit.

For material that can be left at the bottom of the pit, the cut-off grade is x_c such that the cost of mining and processing is equal to zero:

$$x_c \cdot 0.859 \cdot (1.00 - 0.30) \cdot 2{,}205 - 1.00 - 3.00 = 0.00$$

$$x_c = 0.302\%\text{Cu}$$

OPPORTUNITY COSTS AND BENEFITS

Opportunity costs or benefits, $U_{opp}(x)$, may result from mining and processing one metric ton of material not previously scheduled for processing. No opportunity cost is incurred if the mine, mill, and refining facilities are not capacity constrained and if adding one more metric ton to the process has no impact on previously expected cash flows. If there is a capacity constraint, the opportunity cost includes the cost of displacing material already scheduled for processing and postponing treatment of this material.

Capacity Constraints and Opportunity Costs

Consider a project for which the net present value of future cash flows (NPV) was calculated on the basis of currently planned production. According to the current plan, the processing plant has no spare capacity. If one new metric ton of material is added to the capacity-constrained processing plant, treatment of the originally scheduled material is postponed by the time needed to process the additional metric ton. Processing one metric ton of material takes t units of time, and adding one new metric ton today will decrease the net present value of future cash flows by $t \cdot i \cdot NPV$, where i is the discount rate used to calculate the net present value. Therefore, the opportunity cost of adding one metric ton of material to a capacity-constrained operation can be calculated as follows:

$$U_{opp}(x) = -t \cdot i \cdot NPV$$

The opportunity cost must be added to the direct cost of the process that is capacity limited. If one new metric ton of ore is sent to a capacity-constrained mill, t is the time needed to mill this metric ton, and the opportunity cost must be added to the processing cost P. If the refining process is capacity bound, t is the time needed to refine the concentrate produced from one metric ton of material at grade x, and the opportunity cost must be added to the refining cost R.

Constraints on Mining or Processing Capacity: Precious Metal Example

Consider an underground gold mine for which the net present value of future cash flows has been calculated at 100 million dollars (NPV = $100,000,000) using a 15% discount rate (i = 15%). The mine shaft is capacity constrained, with a maximum haulage capacity of 2 million (2,000,000) metric tons per year. Consideration is being given to mining low-grade material on the periphery of high-grade stopes. The time needed to mine and deliver to the surface one metric ton of material is $t = 1/2{,}000{,}000$ year. The opportunity cost of adding one new metric ton to the production schedule can be calculated as follows:

$$U_{opp}(x) = -15\% \cdot \$100{,}000{,}000/2{,}000{,}000$$
$$= -\$7.50 \text{ per metric ton of ore mined}$$

Assume that the following parameters apply to ore being processed: r = 90%, V = $270.00 per ounce of gold, R = $5.00 per ounce, M = $40.00 per metric ton mined and processed, P = $20.00 per metric ton processed, and O = 20% of operating costs. If only direct costs and revenues are taken into account, the cut-off grade between ore and waste can be determined as follows:

$$x_c = [1.20 \cdot (40.00 + 20.00)]/[0.90 \cdot (270.00 - 5.00)]$$
$$x_c = 0.302 \text{ ounce/metric ton} = 9.39 \text{ grams/metric ton}$$

When adding the $7.50 opportunity cost to the mining cost, the cut-off grade is increased by nearly one gram per metric ton:

$$x_c = [1.20 \cdot (40.00 + 20.00) + 7.50]/[0.90 \cdot (270.00 - 5.00)]$$
$$x_c = 0.333 \text{ ounce/metric ton} = 10.37 \text{ grams/metric ton}$$

When the mine approaches the end of its economic life, the net present value of future cash flows decreases toward zero and so does $U_{opp}(x)$. In the preceding example, the cut-off grade decreases from 10.37 grams/metric ton at the beginning of the mine life to 9.39 grams/metric ton at the end.

To illustrate the relationship between cut-off grade and year when the ore is mined, assume that the mine discussed previously has a remaining life of 15 years and a net revenue of $14.9 million per year. In year 1, when 15 years of production remain, the project NPV is $100 million, the opportunity cost is $7.50 per metric ton mined, and the optimal cut-off grade is 10.37 grams/metric ton. In year 2, the mine life is reduced to 14 years, the NPV is $97.9 million, the opportunity cost is $7.34, and the cut-off grade is 10.35 grams/metric ton. At the end of the mine life, the NPV is zero and the cut-off grade is 9.39 grams/metric ton.

The relationship between NPV, opportunity cost, and year when the ore is mined is shown in Figure 2-2. According to this relationship, declining cut-off grades must be used to maximize net present value. Figure 2-3 shows the relationship between optimal cut-off grade and year when the ore is mined.

In the preceding example, the opportunity cost resulted from a haulage capacity constraint and is applicable to both waste and ore haulage costs. The cut-off grades shown in Figure 2-3 must only be used to decide which material should be left underground as opposed to being mined and processed. These cut-off grades must be used to determine which stopes should be mined and the size of these stopes. Because the opportunity cost for hauling ore is the

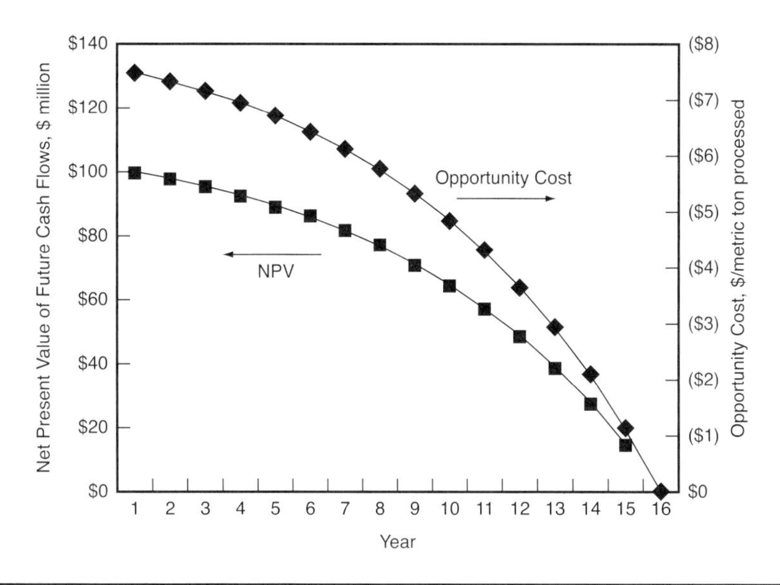

FIGURE 2-2 Relationship between NPV, opportunity cost, and year when the ore is mined

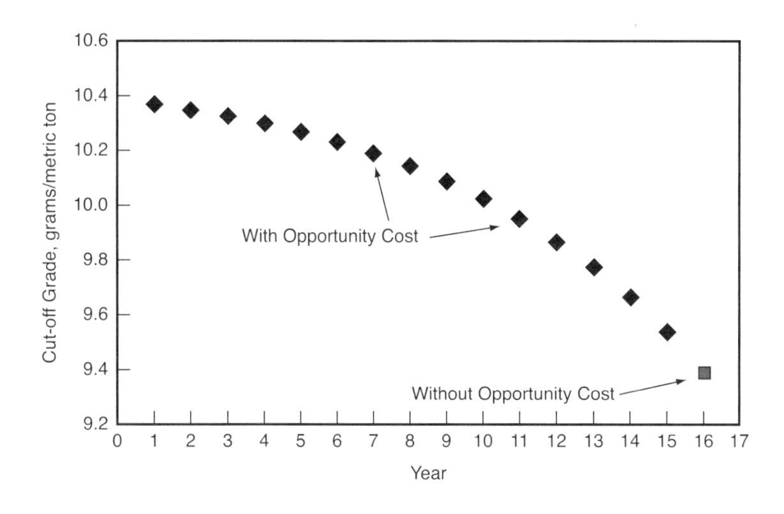

FIGURE 2-3 Relationship between cut-off grade and year when the ore is mined

same as that for hauling the same number of metric tons of waste, the opportunity cost has no bearing on deciding whether material should be processed when it has already been hauled to the surface. For such material, the cut-off grade between ore and waste is independent of the haulage constraint and resulting opportunity cost.

If the processing plant was capacity constrained instead of the mine shaft, the corresponding opportunity cost would apply to all metric tons sent to the mill but not to metric tons wasted. This opportunity cost would enter in all cut-off grade calculations, whether the material was underground or already at the surface. All cut-off grades would be increased accordingly.

Constraints on Smelter Capacity or Volume of Sales: Precious Metal Example

Consider the same gold mining operation described previously, with the new assumption that constraints on mine and plant capacities have been removed, but production constraints are now imposed by the refinery. The refinery can process no more than 600,000 ounces of gold per year, and this capacity is fully utilized. If the cut-off grade is changed to such an extent that one ounce of additional gold is sent to the refinery, the time needed to refine this gold will be $t = 1/600{,}000$ year. With the project's NPV at $100,000,000 and the discount rate at 15%, the opportunity cost of adding one more ounce to the production schedule can be calculated:

$$U_{opp}(x) = -15\% \cdot \$100{,}000{,}000/600{,}000 = -\$25.00 \text{ per ounce}$$

This cost must be added to the refining cost, R = $5.00 per ounce. If there were no capacity constraint, the cut-off grade would be calculated as follows:

$$x_c = [1.20 \cdot (40.00 + 20.00)]/[0.90 \cdot (270.00 - 5.00)]$$

$$x_c = 0.302 \text{ ounce/metric ton} = 9.39 \text{ grams/metric ton}$$

Once the constraint on refining capacity is taken into account, this cut-off grade becomes

$$x_c = [1.20 \cdot (40.00 + 20.00)]/[0.90 \cdot (270.00 - 5.00 - 25.00)]$$

$$x_c = 0.333 \text{ ounce/metric ton} = 10.37 \text{ grams/metric ton}$$

The same formulae should be used if the limit on ounces produced is imposed by marketing constraints, including sales contracts. The opportunity cost must be deducted from the unit value of the product sold.

Constraints on Mining, Milling, or Refining Capacity: Base Metal Example

Consider a copper mining operation characterized by a total mining capacity of 72 million metric tons per year, including both ore and waste. The mill capacity is 36 million metric tons of ore per year and the refining capacity is 299 million pounds per year. At the time the cut-off grade is being calculated, the net present value of future cash flows has been estimated at $300 million using a 10% discount rate. The copper recovery is estimated at 85.9% (including 89% from the flotation plant and 96.5% from the smelter). Freight and smelting costs are $0.30 per pound of copper. The copper price is $1.20 per pound.

$$\begin{aligned} NPV &= \$300{,}000{,}000 \\ i &= 10\% \\ r &= 85.9\% \\ V &= \$1.20 \text{ per pound of copper} \\ R &= \$0.30 \text{ per pound of copper} \end{aligned}$$

Because there are 2,205 pounds in one metric ton, the value of copper contained in one metric ton of material of grade x is calculated as follows:

$$x \cdot r \cdot (V - R) = x \cdot 0.859 \cdot (1.20 - 0.30) \cdot 2{,}205 = \$1{,}705 \cdot x$$

For example, if one metric ton of material contains 1% copper, the value of the copper contained is $17.05.

Assume that the mine is capacity constrained, but the mill and refinery have spare capacity. The opportunity cost to be added to the mining costs is calculated as follows:

$$\begin{aligned} U_{opp}(x) &= -t \cdot i \cdot NPV \\ &= -(1/72{,}000{,}000) \cdot 10\% \cdot \$300{,}000{,}000 \\ &= -\$0.42 \text{ per metric ton mined} \end{aligned}$$

This opportunity cost must be added to the mining cost M of all metric tons, ore or waste, that are subject to mine capacity constraint. It does not change the cut-off grade if the metric ton considered must be either mined and wasted or mined and processed. However, it does increase the cut-off grade if a decision must be made between leaving the material in the ground or mining it and sending it to the mill: a $0.42 increase in mining cost per metric ton results in a 0.02%Cu increase in cut-off grade, calculated as follows:

$$x = \$0.42/\$1{,}705 = 0.02\%\text{Cu}$$

Now assume that the mill is capacity constrained, but the mine and refinery are not. The opportunity cost to be added to the processing cost P is

$$U_{opp}(x) = -(1/36{,}000{,}000) \cdot 10\% \cdot \$300{,}000{,}000$$
$$= -\$0.38 \text{ per metric ton processed}$$

All metric tons milled are subject to this increase in cost. The mill cut-off grade must be increased by 0.05%Cu, calculated as follows:

$$x = \$0.38 / \$1{,}705 = 0.05\%\text{Cu}$$

Finally, assume that the refinery is capacity constrained, but the mine and mill are not. The opportunity cost to be added to the refining cost R is

$$U_{opp}(x) = -(1/299{,}000{,}000) \cdot 10\% \cdot \$300{,}000{,}000$$
$$= -\$0.10 \text{ per pound of copper}$$

When taking this opportunity cost into account, the value of the copper contained in one metric ton of ore of average grade x is reduced from $\$1{,}705 \cdot x$ (as calculated previously) to

$$x \cdot r \cdot [V - R - U_{opp}(x)] = x \cdot 0.859 \cdot (1.20 - 0.30 - 0.10) \cdot 2{,}205$$
$$= \$1{,}515 \cdot x$$

To compensate for this decrease in value, the cut-off grade must be increased by 12.5% calculated as follows: \$1,705 / \$1,515 = 12.5%.

CUT-OFF GRADE OPTIMIZATION WITH OPPORTUNITY COSTS

The formula $U_{opp}(x) = -t \cdot i \cdot NPV$ is useful to verify that cut-off grades and NPV have been optimized. However, cut-off grades calculated from cash flows that have not been optimized are also not optimal and an iterative approach must be used. For example, one could first calculate a cash flow using a fixed cut-off grade, such as that calculated without opportunity cost. From this cash flow, cut-off grades could be re-estimated using opportunity costs. But new cut-off grades imply new mine plans, new cash flows, and therefore new opportunity costs, which must be used to re-estimate the cut-off grades once again. This iterative process must be repeated until cut-off grades and cash flows converge toward stable values.

This iterative approach to cut-off grade optimization can be a lengthy process. Algorithms can be found in the technical literature and computer programs have been developed to facilitate the process. See the bibliography for detailed information on the technical literature available on this process. However, because of the complex relationship between space-dependent geological properties of the deposit, technical constraints that are a function of mining and processing assumptions, and time-dependent variables that define yearly production and cash flows, no simple solution to this difficult optimization problem has yet been found and none can be expected.

Because of the relationship between cut-off grade, mining capacity, processing capacity, mining and processing costs, market value of product sold, and cash flow, all opportunity costs and other costs and benefits likely to result from a change in cut-off grade must be carefully reviewed before the cut-off grade is changed. Declining cut-off grades can maximize net present value but will lower total undiscounted revenues from sales. Increasing the cut-off grade implies wasting low-grade material that could be processed at a profit. Consideration should be given to stockpiling lower-grade material that could be processed at a later date. Ways to determine whether material should be stockpiled or wasted will be discussed later.

Cut-off grades that were estimated to be optimal when the original mine plan was developed must be continuously re-estimated because changes in current and expected costs and prices and mine and mill performance will result in changes in future cash flow and opportunity costs. Maximizing net present value tends to give no value to actions for which the consequences will be felt only at the end of the mine life. For example, actions may have to be taken throughout the life of a project to minimize future costs of reclamation and environmental compliance. The cost of these actions may be significant from an NPV point of view, but the resulting savings that will be incurred at the end of the mine life may have no impact on the NPV. Similarly, stockpiling low-grade material may increase costs throughout the mine life, but revenues resulting from processing these stockpiles will only be realized at the end of the mine life. Maximizing net present value should never be the sole guide to decision making. Other costs and benefits must be taken into account, which are discussed in the following section.

OTHER COSTS AND BENEFITS

Cut-off grades play a critical role in defining tonnages mined and processed, average grade of mill feed, cash flows, mine lifetime, and all major characteristics of a mining operation. In addition to the economically quantifiable financial impact that cut-off grade changes may have, other costs and benefits must be taken into account, although they are often not easily quantifiable. Consideration must be given not only to changes in NPV and cash flow as measured by $U_{dir}(x)$ and $U_{opp}(x)$, but also to all other impacts, $U_{oth}(x)$, including those of an environmental, socio-economic, ethical, or political nature. Costs and benefits to all stakeholders must be evaluated. For most mining operations, the following stakeholders must be taken into account:

- Shareholders, who supply the capital needed for the operation and expect a return on their investment
- Banks, who contribute to the supply of financial resources the mining company needs to operate or expand
- Analysts, who advise the investing community
- Employees and their families
- Users of the final product sold by the mining operation, whether it is coal, gold, copper concentrate, iron ore, processed metal, or industrial minerals
- Suppliers, from whom the mining operation purchases equipment, energy, consumables, supplies, services, or expertise
- Local communities, including neighbors of the mining operation
- The local, regional, federal, or country governments, who are responsible for the welfare of their citizens and benefit from the taxes levied from the mining company. These governments must plan for new infrastructure, roads, health, education, and entertainment; increases in traffic, crime, and prostitution; and higher demand for water, food, and housing. They also have a fiduciary duty to ensure appropriate exploitation of national resources.
- Future generations that will live with the long-term impact, good or bad, of the mining operation
- Non-governmental organizations whose mission, self-appointed or otherwise, is to defend the interests of some of the listed stakeholders

Senior management decides how to balance the needs, interests, and requirements of the different stakeholders. Those in charge of mine planning must be given practical guidelines, including guidelines for cut-off grade determination, to ensure that the projects are designed to reach the company's objectives. Maximizing shareholder value (including minimizing shareholder liability) is often quoted as a company's primary objective. However, a company's objectives must include recognition of responsibilities toward all stakeholders, not only the shareholders.

Higher cut-off grades may increase short-term profitability and enhance return to shareholders and other financial stakeholders. Higher cut-off grades may shorten the payback period, thus reducing political risk of creeping or outright nationalization. But reduced mine life reduces time-dependent opportunities, such as those offered by price cycles. Conversely, lower cut-off grades may increase project life with longer economic benefit to all stakeholders, including shareholders, employees, local communities, and government. Longer mine life may result in more stable employment, less socio-economic disruption to local communities, and more stable tax revenues to government. Lower cut-off grades imply fuller consumption of mineral resources, which may present political advantages or may be required by law. All stakeholders may have to choose between higher financial returns over short time periods or lower returns over longer time periods. Using high but decreasing cut-off grades early in the mine life and stockpiling low-grade material for later processing can help balance financial returns and mine life.

One method of optimizing cut-off grades while taking into account unquantifiable costs and benefits consists of evaluating the project under a variety of constraints imposed on discount rate, mine or mill capacity, volume of sales, capital or operating costs, and so forth. Changes in the opportunity cost of imposing these constraints $U_{opp}(x)$ are compared with the corresponding changes in other costs $U_{oth}(x)$. The optimal cut-off grade is that for which the marginal (and quantifiable) increase in opportunity cost is equal to the corresponding marginal (but subjective) decrease in other costs.

CHAPTER THREE

Minimum Cut-off Grades

Minimum cut-off grades are those that apply to situations in which only direct operating costs are taken into account. Capacity constraints are ignored. Cash flows are not discounted. Opportunity costs are not taken into consideration and neither are other consequences, financial or otherwise, that changing the cut-off grade may have on mining and processing plans and cash flows.

CUT-OFF GRADE BETWEEN ORE AND WASTE

Consider material for which the decision has already been made that it will be mined, so the remaining question is whether it should be sent to the processing plant or wasted.

Mathematical Formulation

Using notations introduced previously, the utility of mining and processing one metric ton of ore grade material can be written as follows:

$$U_{ore}(x) = x \cdot r \cdot (V - R) - (M_o + P_o + O_o)$$

x = average grade
r = recovery, or proportion of valuable product recovered from the mined material
V = value of one unit of valuable product
R = refining costs, defined as costs that are related to the unit of valuable material produced
M_o = mining cost per metric ton of ore
P_o = processing cost per metric ton of ore
O_o = overhead cost per metric ton of ore

The utility of mining and wasting one metric ton of waste material can be written as follows:

$$U_{waste}(x) = -(M_w + P_w + O_w)$$

M_w = mining cost per metric ton of waste
P_w = processing cost per metric ton of waste, as may be needed to avoid potential water contamination and acid generation
O_w = overhead cost per metric ton of waste

The minimum cut-off grade is the value x_c of x for which

$$U_{ore}(x_c) = U_{waste}(x_c)$$
$$x_c = [(M_o - M_w) + (P_o - P_w) + (O_o - O_w)]/[r \cdot (V - R)]$$

In this formula, the numerator represents the difference between mining, processing, and overhead costs incurred when treating the material as ore and those incurred when treating the same material as waste. In the denominator, the metal recovery r must be that which applies to material of grade x_c, which is not necessarily equal to the average recovery for all material sent to the processing plant.

This cut-off grade applies to material that must be mined and is sometimes called "internal cut-off grade," as it is that which applies to one metric ton of material located within the limits of an open pit mine or an underground stope.

If the costs of mining and shipping material to the waste dump or to the primary crusher are the same ($M_o = M_w$) and there are no significant additional costs in processing waste ($P_w = 0$ and $O_w = 0$), this cut-off grade is only a function of mill costs and recoveries and is independent of mining costs:

$$x_c = [P_o + O_o]/[r \cdot (V - R)]$$

This cut-off grade being independent of mining costs is sometimes called "mill cut-off grade." 1

Precious Metal Example

As an example, consider a gold oxide leaching operation in which the cost (including overhead) of hauling material to the leach pad is $1.20 and that of sending it to the waste dump is $1.00. The leaching cost, including the cost of producing doré from solution, incremental cost of leach pad expansion, and overhead cost, is $2.00 per metric ton placed. The gold recovery, including

1 The cut-off grade that applies to material that does not have to be mined but can be left at the bottom of an open pit mine or in the walls of an underground mine is sometimes called *mine cut-off grade*.

leaching, processing, and refining recoveries, is 60%. The revenue expected from the sale of recoverable gold in doré is \$5.00 less than the London Metal Exchange gold price. Assuming a \$270.00 per ounce gold price, the minimum cut-off grade is calculated as follows:

$$x_c = [(1.20 - 1.00) + 2.00]/[0.60 \cdot (270.00 - 5.00)]$$

$$x_c = 0.014 \text{ ounce per metric ton} = 0.43 \text{ gram/metric ton}$$

A conversion factor of 31.1035 grams per troy ounce was used in this calculation. A graphical representation of the relationship between $U_{ore}(x)$, $U_{waste}(x)$, and the grade x is shown in Figure 3-1, where x is expressed in grams per metric ton and U(x) in dollars:

$$U_{ore}(x) = 0.60 \cdot (270.00 - 5.00) \cdot x/31.1035 - 1.20 - 2.00$$

$$= 5.112x - 3.20$$

$$U_{waste}(x) = -1.00$$

Material of grade x is wasted or treated as ore depending on which one of the two lines, $U_{waste}(x)$ or $U_{ore}(x)$, is highest on the graph. The cut-off grade is the value x_c of x where both lines intersect: $x_c = 0.43$ gram/metric ton. Leaching material for which the average grade is between 0.43 gram/metric ton and 0.63 gram/metric ton results in a loss, but this loss is less than the cost of sending the same material to the waste dump.

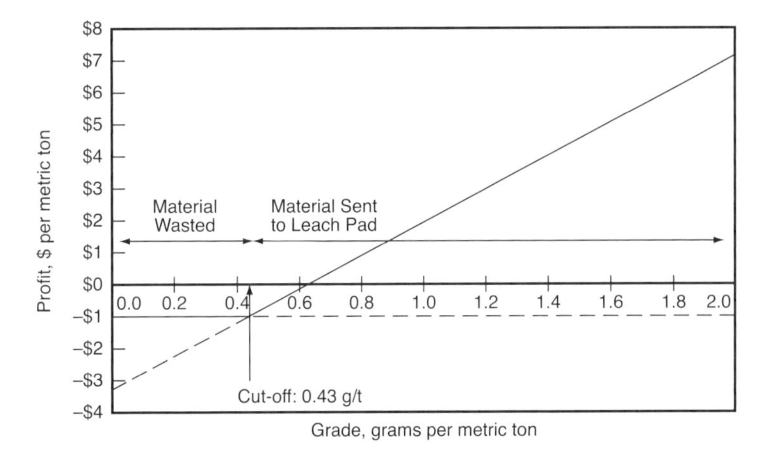

FIGURE 3-1 Graphical estimation of cut-off grade between waste and leached material for material within pit limits

Base Metal Example

Consider a copper mine characterized as follows:

$r =$ 85.9% (including 89% mill recovery and 96.5% smelter recovery)
$V =$ \$1.20 per pound of copper sold
$R =$ \$0.30 per pound of copper for freight, smelting, and refining
$M_o =$ \$1.00 per metric ton of ore mined
$P_o =$ \$3.00 per metric ton of ore processed
$O_o =$ \$0.50 per metric ton of ore processed
$M_w =$ \$1.00 per metric ton of waste
$P_w =$ \$0.05 per metric ton of waste
$O_w =$ \$0.05 per metric ton of waste

The cut-off grade applicable to one metric ton of material that must be mined and can be either processed or wasted (*mill cut-off grade*) is

$$\begin{aligned} x_c &= [(M_o - M_w) + (P_o - P_w) + (O_o - O_w)]/[r \cdot (V - R)] \\ &= \frac{[(1.00 - 1.00) + (3.00 - 0.05) + (0.50 - 0.05)]}{[0.859 \cdot (1.2 - 0.30) \cdot 2{,}205]} \\ &= 0.20\%\text{Cu} \end{aligned}$$

CUT-OFF GRADE FOR MATERIAL AT THE BOTTOM OF AN OPEN PIT MINE

Now consider an open pit mine that is reaching the end of its life. Material is exposed at the bottom of the pit that need not be mined. Alternatively, this material could be mined and processed. What cut-off grade should be used to decide between these two options?

Mathematical Formulation

Because material exposed at the bottom of the pit need not be mined, the utility of leaving it at the bottom of the pit is zero: $U_{waste}(x) = 0$. It should be mined only if it can be both mined and processed at a profit: $U_{ore}(x) > 0$. For such material, the minimum cut-off grade is that which satisfies the following equation:

$$U_{ore}(x_c) = 0$$

$$x_c = [M_o + P_o + O_o]/[r \cdot (V - R)]$$

In this formula, mining, processing, and overhead costs that apply to material remaining at the bottom of the pit may be higher or lower than those prevailing when the mine was at full capacity. This cut-off grade is sometimes called *breakeven cut-off grade* or *mine cut-off grade*.

Precious Metal Example

Consider a gold leaching operation in which mining costs M_o and processing costs P_o, including overhead costs O_o, are \$1.20 and \$2.00, respectively. Gold recovery is 60%, the gold price is \$270.00 per ounce, and a \$5.00 per ounce deduction must be made for shipping, refining, and other charges. The utility of sending material to the leach pad is

$$\begin{aligned} U_{ore}(x) &= x \cdot r \cdot (V - R) - (M_o + P_o + O_o) \\ &= x \cdot 0.60 \cdot (270.00 - 5.00) - (1.20 + 2.00) \\ &= 159x - 3.20 \end{aligned}$$

The utility of leaving material in the pit is

$$U_{waste}(x) = 0$$

The minimum grade at which material located at the bottom of the pit can be mined at a profit is

$$x_c = 3.20/159 = 0.020 \text{ ounce per metric ton} = 0.63 \text{ gram/metric ton}$$

The utility of sending material to the leach pad $U_{ore}(x)$ and that of leaving the material at the bottom of the pit are plotted on Figure 3-2 as a function of the grade x.

Base Metal Example

Consider a copper mine in which mining costs are \$1.00 per metric ton, processing costs are \$3.00 per metric ton, and overhead costs are \$0.50 per metric ton. The copper recovery is 85.9%. The copper price is \$1.20 per pound, from which must be deducted miscellaneous charges amounting to \$0.30 per pound. Prices and costs that are specified in dollars per pound must be converted to dollars per metric ton, taking into account the 2,205-pounds-per-metric-ton conversion factor. The corresponding mine cut-off grade is calculated as follows:

$$\begin{aligned} x_c &= [M_o + P_o + O_o]/[r \cdot (V - R)] \\ &= [1.00 + 3.00 + 0.50]/[0.859 \cdot (1.20 - 0.30) \cdot 2{,}205] \\ &= 0.26\%\text{Cu} \end{aligned}$$

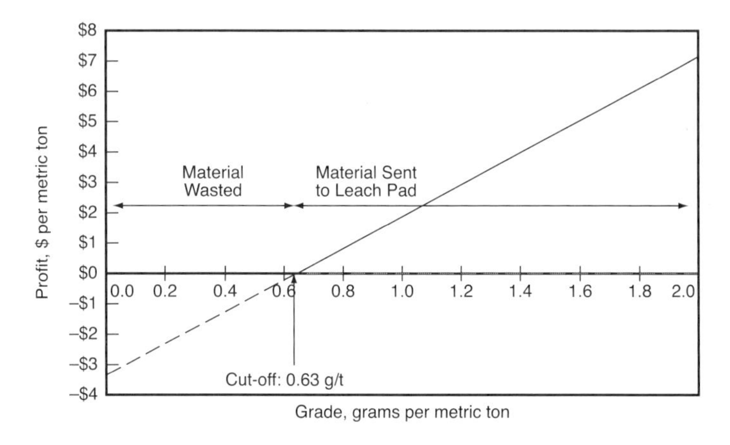

FIGURE 3-2 Graphical estimation of cut-off grade between waste and leached material for material at the bottom of the pit

This mine cut-off grade separates material that can be left in situ from that which can be processed. It can be compared with the 0.20%Cu mill cut-off grade calculated previously.

CUT-OFF GRADES IN UNDERGROUND MINES

Capacity constraints are common in underground mines. These may include constraints imposed by ore body geometry, geotechnical conditions, shaft and haulage capacities, ventilation requirements, mining method, size and type of mining equipment, health and safety regulations, and other constraints that limit production from a stope, a mine section, or the mine as a whole.

A minimum grade is occasionally quoted when referring to the average grade that a stope must exceed before it is considered for mining. Strictly speaking, this is not a cut-off grade but an average grade, which must be linked to a tonnage. The minimum stope average grade depends on the size of the stope, its location with respect to existing facilities, ease of access, and other stope-specific characteristics. This average grade is that for which the cost of developing the stope and mining it is expected to be less than the profit made by processing the ore and selling the final product. This calculation must be made on a discounted basis, taking all physical constraints into account.

When designing a stope, one must take into account constraints imposed by mining method and geotechnical conditions. One must also determine whether lower-grade material located along the boundary of the stope should be included in the stope. Such material should be mined only if the expected value of the recoverable product it contains exceeds all incremental costs, including mining, haulage, processing, backfilling, and other costs. The minimum cut-off grade that defines boundary material that should be mined is the mine cut-off grade and is estimated using a formula similar to that for material at the bottom of an open pit mine:

$$x_c = [M_o + P_o + O_o]/[r \cdot (V - R)]$$

As an example, consider an underground gold mine where the incremental mining cost is $40.00 per metric ton, the mill processing cost is $20.00 per metric ton, and the mill recovery is 95%. Given a gold price of $270 per ounce and a refining cost of $5.00 per ounce, the minimum cut-off grade to be considered to design a stope can be calculated as follows:

$$x_c = [40.00 + 20.00]/[0.95 \cdot (270.00 - 5.00)]$$

$$x_c = 0.238 \text{ ounce/metric ton} = 7.40 \text{ grams/metric ton}$$

This cut-off grade applies not only to lower-grade material surrounding a high-grade core but also to diluted material (mixture of ore and waste material), which might have to be mined to design physically achievable stope boundaries. Both planned and unplanned dilution must be taken into account. Opportunity costs, such as those imposed by haulage capacity, should be taken into account, which will increase the cut-off grade.

If low-grade material must be mined because it is located within a stope or within other planned openings such as shafts, drifts, crosscuts, and so forth, a lower cut-off grade should be used to determine whether this material should be wasted or processed. For such material, blasting and haulage costs must be incurred whether the material is treated as ore or waste. Only incremental costs need be considered. The minimum cut-off grade is estimated using the formula presented previously for material in the middle of an open pit mine:

$$x_c = [(M_o - M_w) + (P_o - P_w) + (O_o - O_w)]/[r \cdot (V - R)]$$

If ore and waste mining costs are the same ($M_o = M_w$) and waste processing and overhead costs are negligible ($P_w = 0$ and $O_w = 0$), this formula can be written

$$x_c = [P_o + O_o]/[r \cdot (V - R)]$$

The mill cut-off grade is recognized here.

Applicable opportunity costs, which in this case are likely to be only those imposed by mill capacity constraints, should also be taken into account.

CUT-OFF GRADE TO CHOOSE BETWEEN PROCESSES

If two processes are available to treat the same material, cut-off grades must be calculated to separate waste from ore being processed and to decide to which one of the two processes the ore should be sent. How to decide whether material should be processed or wasted was discussed previously.

Mathematical Formulation

To decide between two processes, the utility of sending material of grade x to process 1 must be compared with that of sending the same material to process 2. Mining costs, including haulage cost to the processing plant, may vary depending on the process. Processing costs will be different and so will metallurgical recoveries and overhead costs. If the product sold is a function of the process being used, even the revenue per metric ton produced may differ. The cut-off grade between two processes is calculated using the following formulae, in which subscripts refer to process number:

$$U_1(x) = x \cdot r_1 \cdot (V - R_1) - (M_{o1} + P_{o1} + O_{o1})$$

$$U_2(x) = x \cdot r_2 \cdot (V - R_2) - (M_{o2} + P_{o2} + O_{o2})$$

$$U_1(x_c) = U_2(x_c)$$

$$x_c = \frac{[(M_{o1} - M_{o2}) + (P_{o1} - P_{o2}) + (O_{o1} - O_{o2})]}{[r_1 \cdot (V - R_1) - r_2 \cdot (V - R_2)]}$$

Precious Metal Example

Consider a gold mine where two processing facilities are available: a leach plant for which the processing cost is $2.00 per metric ton and recovery is 60%, and a mill for which the processing cost is $12.00 per metric ton and recovery is 90%. The gold price is $270.00 per ounce, from which must be deducted a $5.00-per-ounce charge. Assuming no capacity constraint and that all other costs are the same, the cut-off grade between the two facilities is

$$x_c = [12.00 - 2.00]/[(0.90 - 0.60) \cdot (270.00 - 5.00)]$$

$$x_c = 0.126 \text{ ounce per metric ton} = 3.91 \text{ grams/metric ton}$$

A graphical representation of the relationship between cut-off grade, process, and net revenue or loss is shown in Figure 3-3.

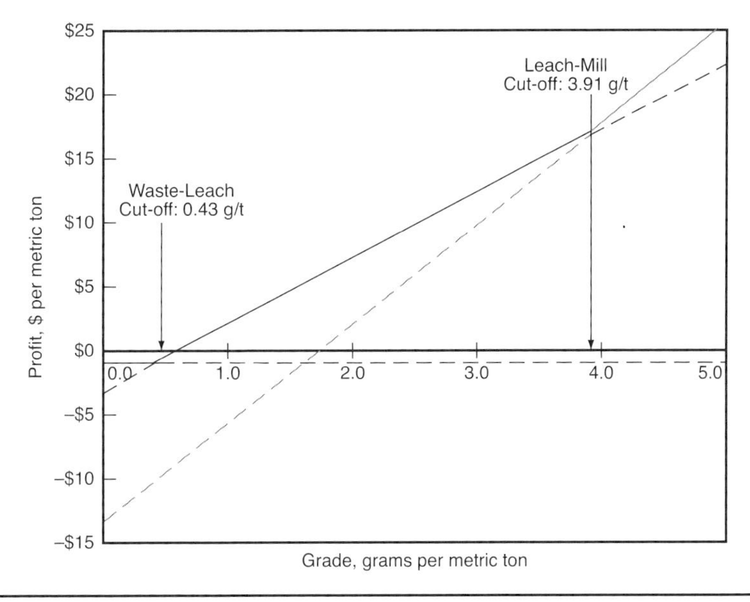

FIGURE 3-3 Graphical estimation of cut-off grade between wasted, leached, and milled material

Base Metal Example

Consider a copper mine for which production can be either leached or milled. The following parameters characterize the conditions under which cut-off grades must be estimated:

r_1 = 85.9% milling and smelting recovery (89% mill, 96.5% smelter)
r_2 = 60.0% average heap leach recovery
V = $1.20 per pound of copper sold
R_1 = $0.30 per pound of copper (including freight and smelting costs of $145.00 per metric ton of concentrate and refining costs of $0.065 per pound of copper)
R_2 = $0.15 per pound of copper for SX-EW and cathode freight to market
M_{o1} = $1.00 mining cost per metric ton of mill ore
M_{o2} = $1.10 mining cost per metric ton of leach ore
P_{o1} = $3.00 processing cost per metric ton of mill ore
P_{o2} = $0.20 processing cost per metric ton of leach ore
O_{o1} = $0.50 overhead cost per metric ton of mill ore
O_{o2} = $0.05 overhead cost per metric ton of leach ore

$M_w =$ \$1.00 mining cost per metric ton of waste
$P_w =$ \$0.05 processing cost per metric ton of waste
$O_w =$ \$0.05 overhead cost per metric ton of waste

Prices and costs that are specified in dollars per pound must be converted to dollars per metric ton, taking into account the 2,205-pounds-per-metric-ton conversion factor. The cut-off grade between leach grade material and mill grade material is

$$x_c = \frac{[(M_{o1} - M_{o2}) + (P_{o1} - P_{o2}) + (O_{o1} - O_{o2})]}{[r_1 \cdot (V - R_1) - r_2 \cdot (V - R_2)]}$$

$$= \frac{[(1.00 - 1.10) + (3.00 - 0.20) + (0.50 - 0.05)]}{[0.859 \cdot (1.20 - 0.30) \cdot 2{,}205 - 0.60 \cdot (1.20 - 0.15) \cdot 2{,}205]}$$

$$= 1.00\%\text{Cu}$$

The cut-off grade between leach grade material and waste is

$$x_c = \frac{[(M_{o2} - M_w) + (P_{o2} - P_w) + (O_{o2} - O_w)]}{[r_2 \cdot (V - R_2)]}$$

$$= \frac{[(1.10 - 1.00) + (0.20 - 0.05) + (0.05 - 0.05)]}{[0.60 \cdot (1.20 - 0.15) \cdot 2{,}205]}$$

$$= 0.02\%\text{Cu}$$

CUT-OFF GRADE BETWEEN WASTE AND LOW-GRADE STOCKPILE

Consideration may be given to stockpiling low-grade material instead of wasting it if such material is not currently economical to process but metal prices are expected to be higher at a later date. Stockpiling low-grade material may also be considered when capacity constraints prevent current processing of material that otherwise could be processed economically. To decide whether material of grade x should be wasted or stockpiled, one must compare the utility of wasting $U_{waste}(x)$ with that of stockpiling $U_{stp}(x)$. The cut-off grade between stockpile and waste is the value x_c of x for which $U_{stp}(x) = U_{waste}(x)$.

The utility of wasting material of grade x can be calculated as follows:

$$U_{waste}(x) = -(M_w + P_w + O_w)$$

To calculate the utility of stockpiling, one must take into consideration stockpiling costs and the cost of retrieving material from stockpile and processing it at a later date. In addition, metallurgical recoveries of stockpiled material may differ from those of freshly mined material, and the price of the product sold may be different from that prevailing when the decision to stockpile is made:

$$U_{stp}(x) = -(M_{stp} + P_{stp} + O_{stp})$$
$$- \text{NPV (future costs of stockpile maintenance)}$$
$$- \text{NPV (future rehandling and processing costs)}$$
$$+ \text{NPV (future revenues from sales)}$$

M_{stp} = current mining costs per metric ton delivered to the low-grade stockpile

P_{stp} = current costs of stockpiling material that will be processed later, including the cost per metric ton of extending the stockpile area if required

O_{stp} = current overhead costs associated with mining and stockpiling

NPV (future costs of stockpile maintenance) = net present value of yearly costs that will be incurred to maintain stockpiled material in an environmentally safe fashion until it is processed

NPV (future rehandling and processing costs) = net present value of the one-time costs that will be incurred when the material is retrieved from the stockpile and processed

NPV (future revenues from sales) = net present value of revenues expected from sales when processed material is sold. At the time of the sale, these revenues will be equal to $x \cdot r_{stp} \cdot (V_{stp} - R_{stp})$:

r_{stp} = recovery expected at the time of processing

V_{stp} = dollar value of the product sold at the time it is sold

R_{stp} = cost per unit of product sold

The recovery r_{stp} may be less or higher than that which would apply to the same material if processed when mined. Sulfide material is likely to oxidize during stockpiling. If a sulfide flotation process is to be used, oxidation will result in lower recovery. Conversely, if an oxide leach process is to be applied to material that was not fully oxidized when mined, stockpiling may enhance recovery.

There are obvious difficulties in using these formulae, the main one being that future costs and revenues are difficult or impossible to estimate with accuracy. Furthermore, because processing of stockpiled material is likely to occur late in the mine life, the net present value of future revenues is likely to be small compared with costs incurred at the time of mining and ongoing maintenance costs during the life of the stockpile. For this reason, stockpiling low-grade material is often a strategic decision that takes into account expectations of future increases in metal prices (V_{stp} could be much higher than V), benefits associated with lengthening the mine life, good management of mineral resources, and other benefits $U_{oth}(x)$ as defined previously in this book.

CUT-OFF GRADE WITH VARIABLE RECOVERIES

General Mathematical Formulae

In the previous examples, it was assumed that the recovery achieved in the processing plant was a constant. For many processes and deposits, the recovery r is a function r(x) of the head grade x. The value of $U_{ore}(x)$ must then be written as follows:

$$U_{ore}(x) = x \cdot r(x) \cdot (V - R) - (M_o + P_o + O_o)$$

The value of $U_{waste}(x)$ remains independent of x:

$$U_{waste}(x) = -(M_w + P_w + O_w)$$

Calculating the cut-off grade requires finding the value of x such that $U_{ore}(x) = U_{waste}(x)$.

Non-linear Recovery: A Precious Metal Example

Consider a gold mine where two processing facilities are available: a leach plant for which the processing cost is $2.00 per metric ton and a mill for which the processing cost is $12.00 per metric ton. Figure 3-4 shows the relationship between recovery and grade, as determined from metallurgical testing and historical production statistics. The gold price is $270.00 per ounce from which must be deducted a $5.00-per-ounce charge.

Figure 3-5 shows the profit that will be made depending on whether material of grade x is wasted ($U_{waste}(x)$), sent to the leach pad ($U_1(x)$), or processed in the mill ($U_2(x)$). It also illustrates how the cut-off grade can be determined by graphical method. The relationship between the utility of leaching or milling material and the average grade of this material is no longer linear. The optimal process for material of grade x is that for which the utility is highest. The cut-off grades are the grades at which the curves intersect. If a constant 60% recovery for leached material and 90% recovery for milled material had been assumed, the ore-leach cut-off would have been estimated at 0.43 gram/metric ton and the leach-mill cut-off at 3.91 grams/metric ton. When variable recoveries are taken into account, the cut-offs are substantially higher, 0.71 gram/metric ton and 5.08 grams/metric ton, respectively.

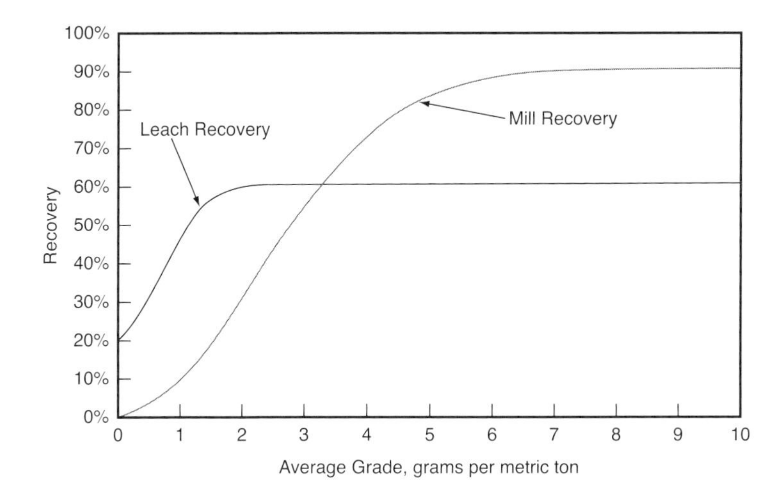

FIGURE 3-4 Relationship between recoveries and average grade

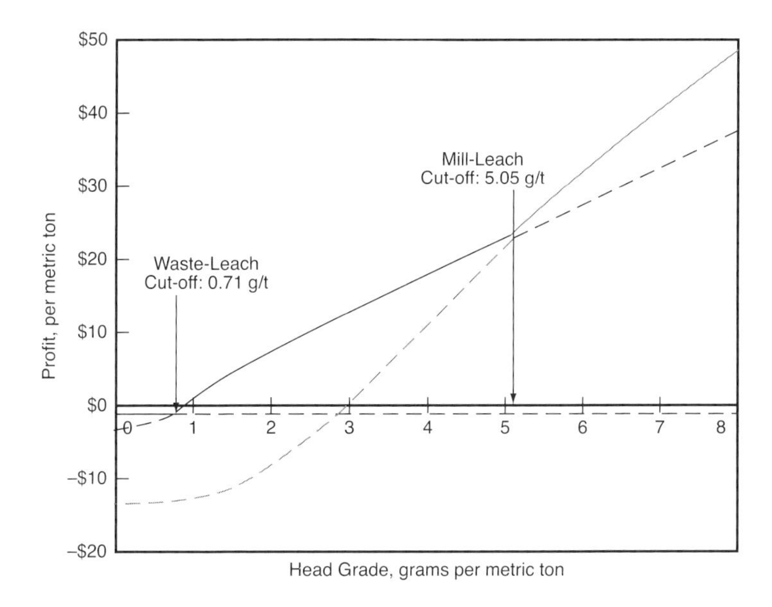

FIGURE 3-5 Graphical estimation of cut-off grade between wasted, leached, and milled material with variable recoveries

Constant Tail: Mathematical Formulation

A model often used to represent the relationship between plant recovery and average grade of plant feed is the constant tail model. This model assumes that a fixed amount of metal cannot be recovered, whatever the grade of the material sent to the plant. If x is the average grade of one metric ton of material and c is the fixed amount that cannot be recovered, the recoverable amount is

$$x \cdot r(x) = r_c \cdot (x - c)$$

x = average grade of material sent to process
$r(x)$ = plant recovery if head grade is x
r_c = constant recovery after subtracting constant tail
c = constant tail

The recovery function is

$$r(x) = r_c \cdot (1 - c/x)$$

Constant Tail: A Base Metal Example

Consider a copper mine characterized as follows:

r_c = 87% (percentage of copper recovered, after deduction of constant tail)
c = 0.04%Cu (constant tail)
V = $1.20 per pound of copper sold
R = $0.30 per pound of copper for freight, smelting, and refining
M_o = $1.00 mining cost per metric ton of ore processed
P_o = $3.00 processing cost per metric ton of ore processed
O_o = $0.50 overhead cost per metric ton of ore processed
M_w = $1.00 mining cost per metric ton of waste
P_w = $0.05 processing cost per metric ton of waste
O_w = $0.05 overhead cost per metric ton of waste

Figure 3-6 illustrates the relationship between recovery r(x) and average grade x.

$$r(x) = 0 \text{ if } x < c$$

$$r(x) = r_c \cdot (1 - c/x) = 0.87 \cdot [1 - 0.04/(100x)] \text{ if } x > c$$

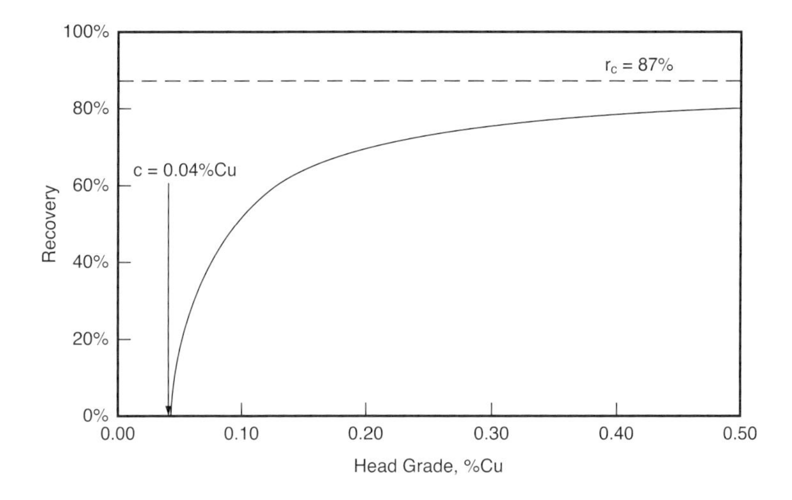

FIGURE 3-6 Relationship between recovery and average grade with constant tail

The relationship between $U_{ore}(x)$, $U_{waste}(x)$, and average grade is shown in Figure 3-7.

$$\begin{aligned}
U_{ore}(x) &= x \cdot r(x) \cdot (V - R) - (M_o + P_o + O_o) \\
&= 0.87 \cdot (x - 0.04/100) \cdot (1.20 - 0.30) \\
&\quad \cdot 2{,}205 - (1.00 + 3.00 + 0.50) \\
&= 1{,}726x - 5.191 \\
U_{waste}(x) &= -(M_w + P_w + O_w) = -(1.00 + 0.05 + 0.05) = -1.10
\end{aligned}$$

The cut-off grade between ore and waste is x_c such that $U_{ore}(x) = U_{waste}(x)$:

$$x_c = 0.24\%\text{Cu}$$

OPPORTUNITY COST OF NOT USING THE OPTIMUM CUT-OFF GRADE

If the optimum cut-off grade is not used, material is sent to a destination where the profit made is less than could be made otherwise or the loss incurred is greater than necessary. Figure 3-8 shows the opportunity cost incurred per metric ton when a leach-mill cut-off grade of 3 grams/metric ton is used although the optimal cut-off grade is 3.91 grams/metric ton. The loss is represented by the

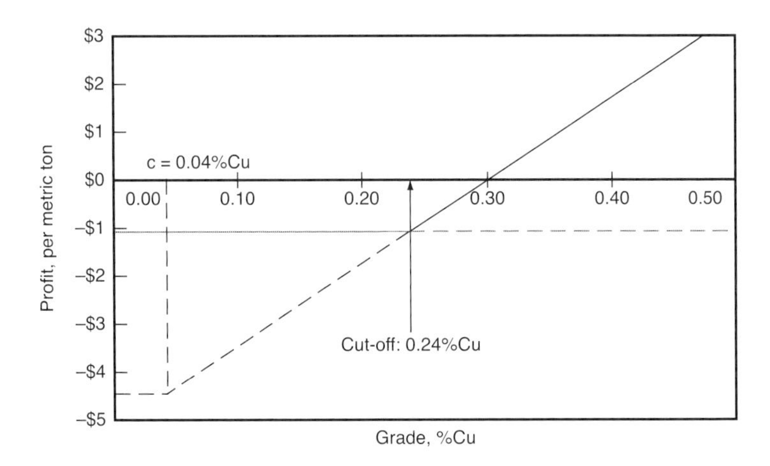

FIGURE 3-7 Graphical estimation of cut-off grade between wasted and milled material with constant tail

difference between the utility of the chosen process and that of the optimal process for the same average grade. Figure 3-9 shows the opportunity cost incurred per metric ton when a leach-mill cut-off grade of 5 grams/metric ton is used.

Let $U_1(x)$ be the utility of leaching one metric ton of material of grade x and $U_2(x)$ the utility of milling the same metric ton. These utilities can be written as follows (in these equations, the cost R is included in V, and the overhead costs O_o are included in M_o, P_{o1}, and P_{o2}):

$$U_1(x) = x \cdot r_1 \cdot V - (M_o + P_{o1})$$
$$U_2(x) = x \cdot r_2 \cdot V - (M_o + P_{o2})$$

The optimal cut-off grade is

$$x_c = (P_{o1} - P_{o2})/[(r_1 - r_2)V]$$

Let x_s be the selected cut-off grade, which is lower than the optimal cut-off grade x_c (Figure 3-8). Material with grade x between x_s and x_c is being milled, which ideally should be leached. For each metric ton of grade x between x_s and x_c, the opportunity cost is

$$U_2(x) - U_1(x) = x \cdot (r_2 - r_1) \cdot V - (P_{o2} - P_{o1})$$

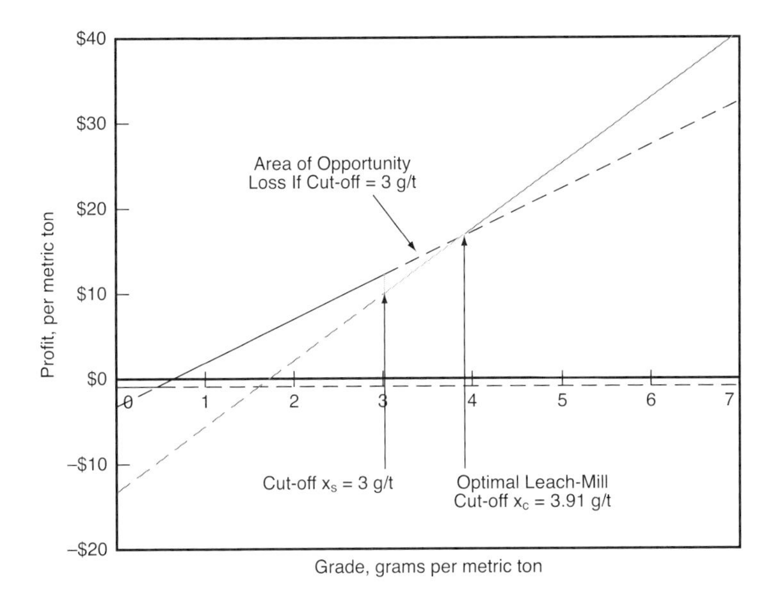

FIGURE 3-8 Opportunity cost of using a cut-off grade lower than the optimal cut-off grade

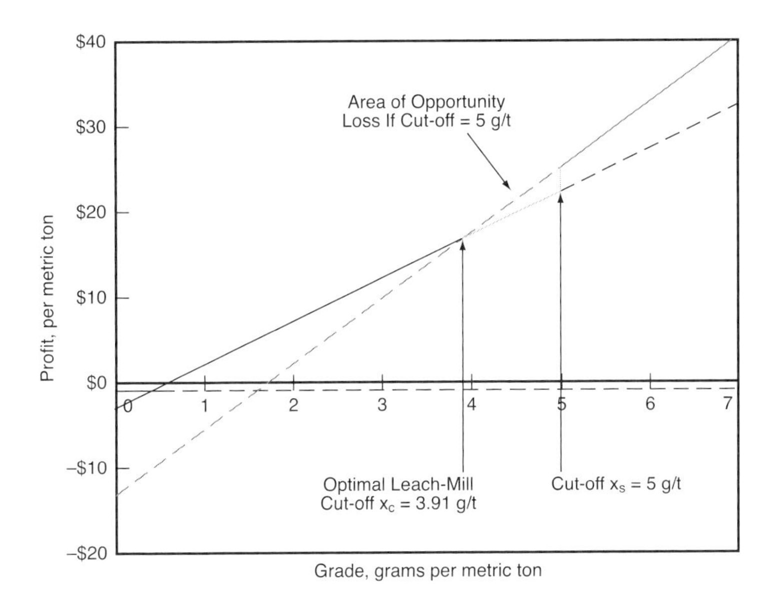

FIGURE 3-9 Opportunity cost of using a cut-off grade higher than the optimal cut-off grade

Integrating this formula from $x = x_s$ to $x = x_c$, the total opportunity cost is obtained:

$$\text{total opportunity cost} = [Q(x_s) - Q(x_c)] \cdot (r_2 - r_1) \cdot V$$
$$-[T(x_s) - T(x_c)] \cdot (P_{o2} - P_{o1})$$

In this formula, $T(x_s) - T(x_c)$ is the tonnage of material with average grade between x_s and x_c and $Q(x_s) - Q(x_c)$ is the quantity of metal contained in this material. It would be possible to avoid this opportunity cost by increasing the mill capacity by a tonnage amount equal to $T(x_s) - T(x_c)$. Such an increase in capacity is justified if the cost of such an increase is expected to be less than the total opportunity cost.

Similar equations are applicable if x_s is higher than x_c and material that should be milled is leached (Figure 3-9):

$$\text{total opportunity cost} = [Q(x_c) - Q(x_s)] \cdot (r_1 - r_2) \cdot V$$
$$-[T(x_c) - T(x_s)] \cdot (P_{o1} - P_{o2})$$

CHAPTER FOUR

Cut-off Grade for Polymetallic Deposits

Polymetallic deposits are defined as deposits that contain more than one metal of economic value. The formulae that must be used to calculate the utility of sending one metric ton of material to a given destination or process must consider the contribution of each metal. The decision whether one metric ton of material should be wasted or sent to the processing plant can no longer be made on the basis of grade alone. Dollar values must be calculated for each possible process, and the cut-off between ore and waste must be expressed in dollar terms.

GENERAL CONSIDERATIONS

Consider a metric ton of material that contains two valuable metals, copper and gold. Let x_1 and x_2 be the copper and gold grades, respectively. The processing plant consists of crushing, grinding, and flotation circuits. A copper concentrate is produced, which is sold to a smelter. The flotation plant recovery is r_1 for copper and r_2 for gold. Mining, processing, and overhead costs associated with one metric ton of material sent to the flotation plant are M_o, P_o, and O_o, respectively. The corresponding costs per metric ton of waste are M_w, P_w, and O_w. According to the smelter contract, the value received for sale of the concentrate is p_1 = 95% of the value of the copper contained in the concentrate after a deduction, d_1, and p_2 = 99% of the gold contained. Smelter cost deductions are C_s per metric ton of concentrate. The concentration ratio K is the number of metric tons of material that must be processed to produce one metric ton of concentrate. The cost of shipping one metric ton of concentrate to the smelter is C_t. Metal prices are those quoted on the London Metal Exchange, V_1 and V_2 for copper and gold, respectively. Therefore, the value of one metric ton of material sent to the flotation plant is

$$\begin{aligned} U_{ore}(x_1, x_2) = {} & (x_1 r_1 - d_1) p_1 V_1 + (x_2 r_2) p_2 V_2 \\ & - C_s/K - C_t/K - (M_o + P_o + O_o) \end{aligned}$$

If the same metric ton is sent to the waste dump, the corresponding costs are

$$U_{waste} = -(M_w + P_w + O_w)$$

The material should be sent to the processing plant if

$$U_{ore}(x_1, x_2) > U_{waste}$$

These formulae show that many factors enter into the calculation of the cut-off between ore and waste. Processing costs and recoveries are likely to be dependent not only on metal content but also on other geological characteristics such as mineralogy, hardness, clay content, and degree of oxidation, which change depending on the area of the deposit being mined. Smelter contracts heavily penalize concentrates that are found to contain excessive amounts of specified deleterious elements. All these factors must be taken into account when estimating the cut-off value applicable to one metric ton of mineralized material.

Because the value of one metric ton of material is a function of more than one grade, it is no longer meaningful to talk about a "cut-off grade." Historically, this multidimensional problem was reduced to a one-dimensional problem by defining a "metal equivalent." With the advance of computers and the ease of use with which complex mathematical calculations can be made, one now refers to *cut-off values*, which are expressed in dollar terms and require calculation of a *net smelter return*. Net smelter return and metal equivalents are discussed in the following paragraphs.

CALCULATION OF CUT-OFF GRADES USING NET SMELTER RETURN

For polymetallic deposits, the utility of sending one metric ton of material to the smelter is best expressed in terms of net smelter return, or NSR. The net smelter return is defined as the return from sales of concentrates, expressed in dollars per metric ton of ore, excluding mining and processing costs.

Mathematical Formulation

In the previous copper–gold example, the NSR of one metric ton of ore with copper grade x_1 and gold grade x_2 is

$$NSR(x_1, x_2) = (x_1 r_1 - d_1) p_1 V_1 + (x_2 r_2) p_2 V_2 - C_s / K - C_t / K$$

The utility of sending this metric ton of ore to the processing plant is

$$U_{ore}(x_1, x_2) = NSR(x_1, x_2) - (M_o + P_o + O_o)$$

Using NSR values greatly simplifies the calculation of cut-off grades. For example, the NSR cut-off between processing and wasting one metric ton of material of average grades x_1, x_2 is NSR_c, calculated as follows:

$$NSR_c - (M_o + P_o + O_o) = -(M_w + P_w + O_w)$$
$$NSR_c = (M_o + P_o + O_o) - (M_w + P_w + O_w)$$

In polymetallic deposits, cut-offs should not be expressed in terms of minimum metal grade; they should be expressed in terms of minimum NSR.

Calculation of NSR Cut-off: A Copper–Molybdenum Example

Consider a copper–molybdenum mining operation. In this section, the subscript 1 refers to copper and 2 refers to molybdenum. Therefore, x_1 is the copper grade and x_2 is the molybdenum grade. The following parameters characterize the operation:

r_1 = 89% copper flotation plant recovery
p_1 = 96.5% copper smelting recovery
r_2 = 61% molybdenum flotation plant recovery
p_2 = 99% molybdenum roasting recovery
V_1 = $1.20 value of one pound of copper sold
V_2 = $6.50 value of one pound of molybdenum sold
R_1 = $0.065 refining cost per pound of copper
K = 72 metric tons of ore that must be processed to produce one metric ton of concentrate
$C_s + C_t$ = $145.00 smelting and freight costs per metric ton of concentrate
R_2 = $0.95 conversion, roasting, and freight costs per pound of molybdenum
M_o = $1.00 mining cost per metric ton milled
P_{o1} = $3.00 mill processing cost per metric ton milled
P_{o2} = $0.15 incremental molybdenum processing cost per metric ton milled
O_o = $0.50 overhead cost per metric ton milled
M_w = $1.00 mining cost per metric ton wasted
P_w = $0.05 processing cost per metric ton wasted
O_w = $0.05 overhead cost per metric ton wasted

The NSR of one metric ton of material with average grade x_1, x_2 is calculated as follows:

$$\begin{aligned} NSR(x_1, x_2) &= x_1 r_1 p_1 (V_1 - R_1) + x_2 r_2 p_2 (V_2 - R_2) - (C_s + C_t)/K \\ &= 0.89 \cdot 0.965 \cdot (1.20 - 0.065) \cdot 2{,}205 \cdot x_1 \\ &\quad + 0.61 \cdot 0.99 \cdot (6.50 - 0.95) \cdot 2{,}205 \cdot x_2 - 145.00/72 \\ &= 2{,}149 x_1 + 7{,}390 x_2 - 2.016 \end{aligned}$$

Therefore, the NSR value of one metric ton of ore averaging $x_1 = 0.45\%$Cu and $x_2 = 0.035\%$Mo is \$10.24.

For material that must be mined but can be either wasted or processed, the cut-off NSR (*mill* or *internal* cut-off NSR) is NSR_c, calculated as follows:

$$\begin{aligned} NSR_c &= (P_{o1} + P_{o2} - P_w) + (O_o - O_w) + (M_o - M_w) \\ &= (3.00 + 0.15 - 0.05) + (0.50 - 0.05) + (1.00 - 1.00) \\ &= \$3.55 \text{ per metric ton} \end{aligned}$$

For material that need not be mined (*mine* or *external* cut-off NSR), NSR_c is calculated as follows:

$$\begin{aligned} NSR_c &= P_{o1} + P_{o2} + O_o + M_o \\ &= 3.00 + 0.15 + 0.50 + 1.00 \\ &= \$4.65 \text{ per metric ton} \end{aligned}$$

The relationship between NSR_c, x_1, and x_2 is shown in Figure 4-1.

CALCULATION AND REPORTING OF METAL EQUIVALENT

Before computers became widely used, it was common practice to refer to polymetallic deposits in terms of metal equivalent. If a metric ton of material contains two metals, copper and molybdenum, with average grades of x_1 and x_2, respectively, the corresponding copper equivalent is defined as the copper grade x_{1e} that one metric ton must contain to produce the same revenue, assuming no molybdenum.

The revenue generated by mining and processing one metric ton of material with copper grade x_1 and molybdenum grade x_2 is $NSR(x_1, x_2)$. The revenue generated by mining and processing one metric ton of material with copper grade x_{1e} and no molybdenum is $NSR(x_{1e}, 0.0)$. The copper equivalent is obtained by solving the following equation:

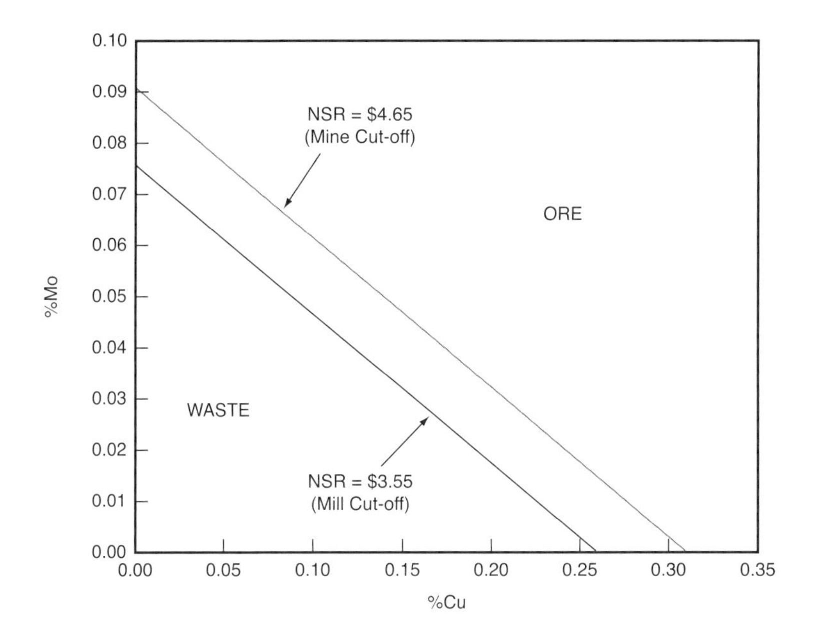

FIGURE 4-1 Relationship between cut-off NSR and metal grades

$$NSR(x_{1e}, 0.0) = NSR(x_1, x_2)$$

A molybdenum equivalent can be calculated instead of a copper equivalent. The molybdenum equivalent is the molybdenum grade x_{2e}, which satisfies the following equation:

$$NSR(0.0, x_{2e}) = NSR(x_1, x_2)$$

In the previous copper–molybdenum example, the NSR was expressed as follows:

$$NSR(x_1, x_2) = x_1 r_1 p_1 (V_1 - R_1) + x_2 r_2 p_2 (V_2 - R_2) - (C_s + C_t)/K$$

Therefore,

$$NSR(x_{1e}, 0.0) = x_{1e} r_1 p_1 (V_1 - R_1) - (C_s + C_t)/K$$

The copper equivalent is

$$x_{1e} = x_1 + x_2 [r_2 p_2 (V_2 - R_2)]/[r_1 p_1 (V_1 - R_1)]$$

Similarly, the molybdenum equivalent is

$$x_{2e} = x_2 + x_1[r_1p_1(V_1 - R_1)]/[r_2p_2(V_2 - R_2)]$$

Using the information listed previously concerning prices, cost, and recoveries, the copper and molybdenum equivalents can be calculated as follows:

$$x_{1e} = x_1 + x_2(7{,}390/2{,}149) = x_1 + 3.439x_2$$

$$x_{2e} = x_2 + x_1(2{,}149/7{,}390) = x_2 + 0.291x_1$$

The copper equivalent of material averaging $x_1 = 0.45\%$Cu and $x_2 = 0.035\%$Mo is 0.57%Cu-equivalent. The molybdenum equivalent of the same material is 0.166%Mo-equivalent.

In practice, because of the complexity of the formulae to be used to estimate the value of a metric ton of material correctly, and because equivalence changes with metal price, recoveries, and refining costs, grade equivalence is rarely a useful tool in calculation of cut-off grades. Quoting the amount of metal equivalent contained in a deposit is of little use to investors. Publication of reserves in terms of metal equivalence is generally not accepted by regulatory agencies unless additional disclosures are made, including publication of the average grade of each metal and explanation of the formula used to calculate metal equivalence.

CHAPTER FIVE

Cut-off Grade and Optimization of Processing Plant Operating Conditions

In this chapter, a method is developed to optimize a copper mining operation where mining capacity is fixed, but the capacity of the processing plant can be changed by changing grind size. Depending on the metallurgical properties of the ore, using a coarser grind will increase plant throughput while reducing cost per metric ton processed and decreasing recovery. Conversely, a finer grind can decrease plant capacity, increase processing cost, and increase recovery.

MATHEMATICAL FORMULATION

The following notations are used in this chapter:

r = processing plant recovery
V = value of copper contained in concentrate, after deduction for smelter loss, and freight, smelting, and refining costs
P_o = cost per metric ton of ore processed, including overhead
x_c = cut-off grade
T_{+c} = tonnage above cut-off grade to be processed in one year
Q_{+c} = quantity of copper to be processed in one year
x_{+c} = average grade above cut-off grade

Because mining operations are fixed, the utility function that must be optimized to estimate the economically optimal grind size is only a function of mill operations and can be written as follows:

$$U(T_{+c}) = Q_{+c} \cdot r(T_{+c}) \cdot V - T_{+c} \cdot P_o(T_{+c})$$

where

$U(T_{+c})$ = utility of running the plant at T_{+c} capacity for one year
$r(T_{+c})$ = processing plant recovery, if plant capacity is T_{+c}
$P_o(T_{+c})$ = cost per metric ton of ore processed, if plant capacity is T_{+c}

Q_{+c} is also a function of T_{+c}. Both Q_{+c} and T_{+c} are functions of the cut-off grade x_c.

The optimal plant capacity is that for which $U(T_{+c})$ reaches a maximum and is calculated by setting the first derivative of $U(T_{+c})$ equal to zero:

$$dU(T_{+c})/dT_{+c} = 0.0$$

$$dU(T_{+c})/dT_{+c} = dQ_{+c}/dT_{+c} \cdot r(T_{+c}) \cdot V - P_o(T_{+c}) + Q_{+c} dr(T_{+c})/dT_{+c} \cdot V - T_{+c} \cdot dP_o(T_{+c})/dT_{+c}$$

If the tonnage processed is changed by a small amount dT_{+c} because of a small change in cut-off grade x_c, the amount of copper contained is increased from $Q_{+c} = T_{+c} \cdot x_{+c}$ to $Q_{+c} + dQ_{+c} = T_{+c} \cdot x_{+c} + dT_{+c} \cdot x_c$. Therefore, $dQ_{+c} = dT_{+c} \cdot x_c$ and the optimal plant capacity is T_{+c} such that

$$x_c \cdot r(T_{+c}) \cdot V - P_o(T_{+c}) + Q_{+c} \cdot dr(T_{+c})/dT_{+c} \cdot V - T_{+c} \cdot dP_o(T_{+c})/dT_{+c} = 0.0$$

If the recovery r and the processing cost P_o were independent of T_{+c}, this equation would be easily solved for x_c:

$$x_c = P_o(T_{+c})/[r(T_{+c}) \cdot V] = P_o/[r \cdot V]$$

The mill cut-off grade is recognized here.

The term $Q_{+c} \cdot dr(T_{+c})/dT_{+c} \cdot V$ represents the change in the value of the product sold in one year that results from the change in recovery. The term $T_{+c} \cdot dP_o(T_{+c})/dT_{+c}$ represents the change in operating cost per year that results from the change in processing cost per metric ton.

In this formulation of the problem, it was assumed that the value V of the product sold is independent of the tonnage processed. This may not be the case if the quality of the concentrate varies with tonnage processed and head grade. It was also assumed that recovery is only a function of tonnage processed and is independent of head grade. More complex equations would apply if these assumptions could not be made.

EXAMPLE: OPTIMIZATION OF GRINDING CIRCUIT IN A COPPER MINE

The following example illustrates how plant capacity can be optimized when mine plans are fixed, no major change can be made to the processing plant, but plant capacity can be increased by changing grinding size. Mine production is fixed for at least one year, and the tonnage, grade, and metal content of copper-bearing material expected to be mined during this one-year period is as shown in Table 5-1 and illustrated in Figure 5-1.

The ore is to be processed in a flotation plant. The mill was designed to operate at the rate of 39.5 million metric tons per year with an average copper recovery of 95%. Under these conditions, the mill's operating costs are $5.24 per metric ton. When mine plans were finalized for the coming year, the expected value of product sold was $1.00 per pound of copper in concentrate, and the following mill cut-off grade was used for planning:

$$x_c = 5.24/(0.95 \cdot 1.00 \cdot 2{,}205) = 0.25\%\text{Cu}$$

As shown in Table 5-1, this cut-off grade implies that the mill feed will be 39.5 million metric tons, averaging 0.381%Cu and containing 332 million

TABLE 5-1 Grade–tonnage relationship for coming year of mining

Cut-off, %Cu	Minable Tonnage, million metric tons	Minable Grade, %Cu	Minable Copper Content	
			thousand metric tons Cu	million pounds Cu
0.15	53.7	0.335	180	397
0.16	52.6	0.340	179	395
0.17	51.4	0.344	177	390
0.18	50.1	0.348	174	384
0.19	48.8	0.352	172	378
0.20	47.5	0.355	168	372
0.21	46.0	0.360	165	365
0.22	44.0	0.365	162	357
0.23	42.8	0.370	159	349
0.24	41.2	0.375	155	341
0.25	39.5	0.381	150	332
0.26	37.7	0.387	146	322
0.27	35.9	0.393	141	311
0.28	34.1	0.399	136	300
0.29	32.1	0.406	131	288
0.30	30.2	0.413	125	275
0.31	28.2	0.421	119	262

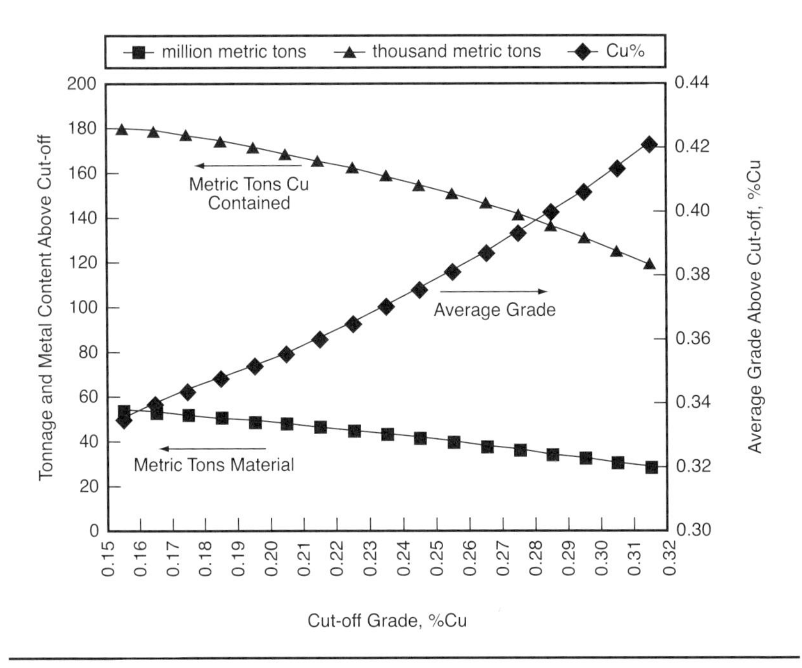

FIGURE 5-1 Graphical representation of grade–tonnage relationship for coming year

pounds of copper. The value of the material sent to the mill, based on \$1.00 per pound of recoverable copper and excluding mining costs, was expected to be

$$
\begin{aligned}
U(T_{+c}) &= Q_{+c} \cdot r(T_{+c}) \cdot V - T_{+c} \cdot P_o(T_{+c}) \\
&= 332 \cdot 0.95 \cdot 1.00 - 39.5 \cdot 5.24 \\
&= \$108 \text{ million}
\end{aligned}
$$

Because of an unexpected increase in copper price, the mining company is investigating whether short-term changes could be made to mill feed and throughput, which would result in increased utility. The copper price is now expected to be \$1.50 per pound of copper in concentrate instead of the \$1.00 that was used for planning. The mine plan cannot be changed for at least one year and only changes in operating conditions can be made to the processing plant. One option is to operate the mine and mill as planned while selling the concentrate at the higher price. The value of the material sent to the mill, excluding mining costs, would increase from \$113 million to

$$
\begin{aligned}
U(T_{+c}) &= 332 \cdot 0.95 \cdot 1.50 - 39.5 \cdot 5.24 \\
&= \$266 \text{ million}
\end{aligned}
$$

Alternatively, one could consider a decrease in cut-off grade. At $1.50 per pound of copper in concentrate, the minimum cut-off grade is

$$x_c = 5.24/(0.95 \cdot 1.50 \cdot 2{,}205) = 0.17\%\text{Cu}$$

Table 5-1 shows that 51.4 million metric tons of ore would be mined above this cut-off grade, averaging 0.344%Cu. Under current operating conditions, the mill can only process 39.5 million metric tons. The higher-grade material could be sent to the mill and the lower-grade material could be stockpiled. But such an approach is likely to increase short-term costs without increasing revenues from concentrate sales. No advantage is taken of the higher copper price.

Another option would consist of increasing mill throughput by increasing grind size. The result would be a decrease in operating cost per metric ton. However, this is expected to result in a decrease in mill recovery. It has been determined that the mill operating costs are 55% fixed costs and 45% inversely proportional to the tonnage processed:

$$P_o(T_{+c}) = 2.88 + 93.1/T_{+c}$$

This relationship between operating cost per metric ton and tonnage processed per year is shown in Figure 5-2.

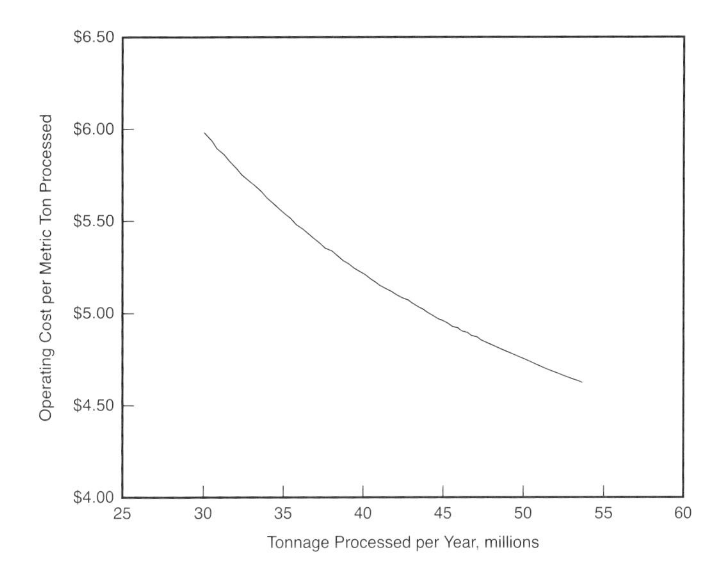

FIGURE 5-2 Relationship between mill operating cost per metric ton and tonnage processed per year

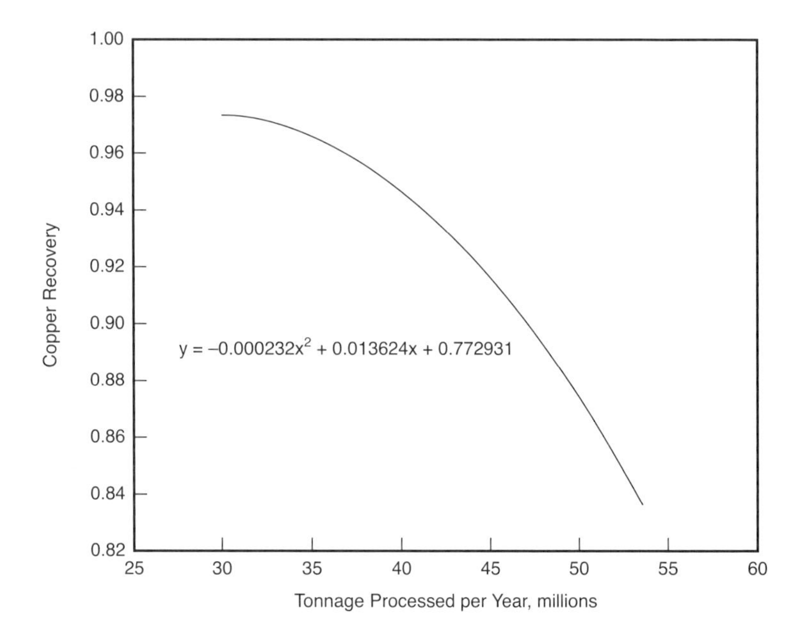

FIGURE 5-3 Relationship between copper recovery and tonnage processed per year

It has also been determined that the relationship between copper recovery and mill throughput is as shown in Figure 5-3. This relationship is represented by the following equation:

$$r(T_{+c}) = -0.000232(T_{+c})^2 + 0.01362T_{+c} + 0.773$$

The function to be optimized is

$$U(T_{+c}) = Q_{+c} \cdot r(T_{+c}) \cdot V - T_{+c} \cdot P_o(T_{+c})$$

The relationship between $U(T_{+c})$ and the cut-off grades (which defines T_{+c}) is easily calculated using Table 5-1 and the two preceding equations. The results are summarized in Table 5-2 and shown in Figure 5-4. The highest return is \$272 million, \$6 million higher than the \$266 million calculated previously when plant capacity was kept at 39.5 million metric tons per year. This highest return is reached by increasing the plant capacity to approximately 45 million metric tons per year.

An alternative method of calculating the optimum processing rate consists of solving the following equation:

TABLE 5-2 Calculation of $U(T_{+c})$ for various cut-off grades and corresponding tonnages of mill feed T_{+c}

Unit of value of Cu in concentrate	V	$/pound	$1.50	$1.50	$1.50	$1.50	$1.50	$1.50
Cut-off grade	x_c	%Cu	0.20%	0.21%	0.22%	0.23%	0.24%	0.25%
Tonnage above cut-off	T_{+c}	million metric tons	47.5	46.0	44.4	42.8	41.2	39.5
Average grade above cut-off	x_{+c}	%Cu	0.355%	0.360%	0.365%	0.370%	0.375%	0.381%
Copper content above cut-off	Q_{+c}	million pounds Cu	372	365	357	349	341	332
Copper recovery	$r(T_{+c})$	%	89.65%	90.86%	92.04%	93.09%	94.03%	94.90%
Unit processing cost	$P_o(T_{+c})$	$/metric ton	$4.84	$4.90	$4.98	$5.06	$5.14	$5.24
Total value of Cu in concentrate	$Q{+c} \cdot r(T_{+c}) \cdot V$	million $/year	$500	$497	$493	$487	$481	$473
Total processing cost	$-T_{+c} \cdot P_o(T_{+c})$	million $/year	$(230)	$(226)	$(221)	$(216)	$(212)	$(207)
Utility	$U(T_{+c})$	million $/year	$270	$272	$272	$271	$269	$266

$$dU(T_{+c})/dT_{+c} = 0.0$$

which can be written as

$$x_c \cdot r(T_{+c}) \cdot V - P_o(T_{+c}) + Q_{+c} \cdot dr(T_{+c})/dT_{+c} \cdot V - T_{+c} \cdot dP_o(T_{+c})/dT_{+c} = 0.0$$

The derivatives of $P_o(T_{+c})$ and $r(T_{+c})$ are easily calculated:

$$dP_o(T_{+c})/dT_{+c} = 93.1/(T_{+c})^2$$

$$dr(T_{+c})/dT_{+c} = -0.000464 \cdot T_{+c} + 0.01362$$

The results obtained are plotted in Figure 5-5 and shown in Table 5-3. The optimal return is obtained if the tonnage of mill feed is set slightly less than 45 million metric tons per year, the point where $dU(T_{+c})/dT_{+c} = 0.0$ (Figure 5-5). Setting the cut-off grade at 0.22%Cu will reach this objective,

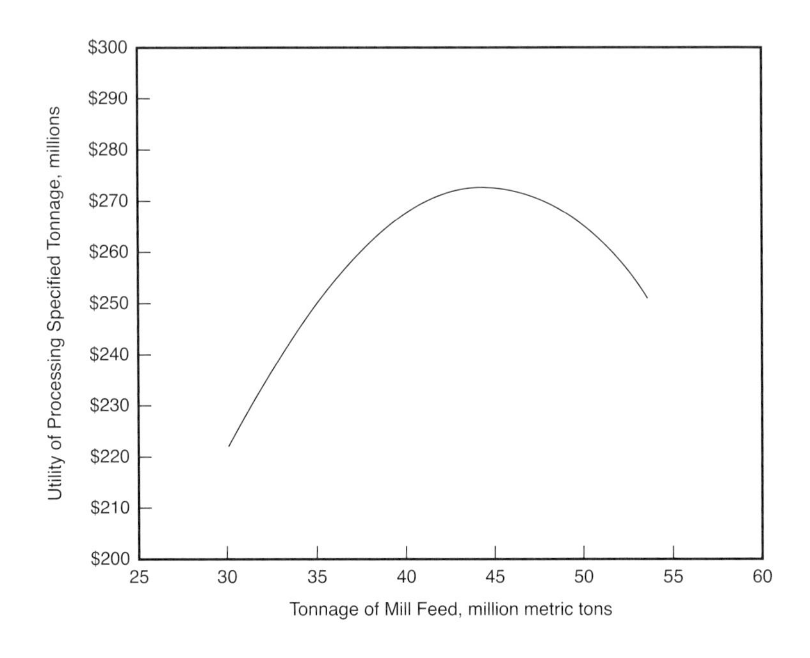

FIGURE 5-4 Relationship between utility $U(T_{+c})$ and tonnage of mill feed T_{+c}

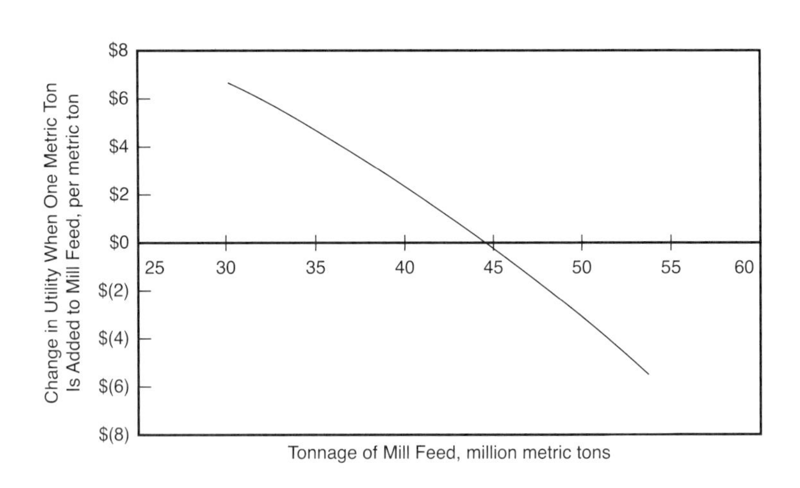

FIGURE 5-5 Relationship between incremental utility $dU(T_{+c})/dT_{+c}$ and tonnage of mill feed T_{+c}

TABLE 5-3 Calculation of $dU(T_{+c})/dT_{+c}$ for various cut-off grades and corresponding tonnages of mill feed T_+

Unit of value of Cu in concentrate	V	$/pound	$1.50	$1.50	$1.50	$1.50	$1.50	$1.50
Cut-off grade	x_c	%Cu	0.20%	0.21%	0.22%	0.23%	0.24%	0.25%
Tonnage above cut-off	T_{+c}	million metric tons	47.5	46.0	44.4	42.8	41.2	39.5
Average grade above cut-off	x_{+c}	%Cu	0.355%	0.360%	0.365%	0.370%	0.375%	0.381%
Copper content above cut-off	Q_{+c}	million pounds Cu	372	365	357	349	341	332
Copper recovery	$r(T_{+c})$	%	89.65%	90.86%	92.04%	93.09%	94.03%	94.90%
Unit processing cost	$P_o(T_{+c})$	$/metric ton	$4.84	$4.90	$4.98	$5.06	$5.14	$5.24
Change in utility when one metric ton is added to mill feed: $dU(T_{+c})dT_{+c}$								
1-if recovery and costs were constant	$x_c \cdot r(T_{+c}) \cdot V - P_o(T_{+c})$	$/metric ton	$1.09	$1.41	$1.72	$2.03	$2.32	$2.61
2-because of change in recovery	$Q_{+c} \cdot dr(T_{+c})/dT_{+c} \cdot V$	$/metric ton	$(4.70)	$(4.23)	$(3.74)	$(3.27)	$(2.81)	$(2.34)
3-because of change in costs	$-T_{+c} \cdot dP_o(T_{+c})/dT_{+c}$	$/metric ton	$1.96	$2.02	$2.10	$2.18	$2.26	$2.36
Utility	$dU(T_{+c})/dT_{+c}$	$/metric ton	$(1.65)	$(0.80)	$0.08	$0.94	$1.77	$2.62

producing 44.4 million metric tons of mill feed. The average mill head grade will be 0.365%Cu. Increasing the tonnage from 39.5 million metric tons to 44.4 million metric tons will be achieved by decreasing recovery from 95% to 92%. This loss in recovery will be more than compensated by a decrease in operating costs from $5.24 to $4.98 per metric ton.

CHAPTER SIX

Cut-off Grade and Mine Planning—Open Pit and Underground Selective Mining

There are many similarities between questions that must be answered when designing open pit and underground mines when selective methods are used. These are illustrated in the examples that follow. Questions that arise in the design of underground bulk mining operations are discussed in the next section.

OPEN PIT MINE: ECONOMIC VALUATION OF A PUSHBACK

Consider the last pushback in an open pit mine. This pushback should be mined only if the net present value (NPV) of the cash flow generated by mining it is positive. The NPV is calculated from the value of each block included in the pushback and can be expressed as follows:

$$NPV = \sum U_{jk}/(1+i)^k$$

$$U_{jk} = \text{utility of mining block j in year k}$$

$$i = \text{discount rate}$$

$$U_{jk} = U_{jk,\,dir} + U_{jk,\,opp} + U_{jk,\,oth}$$

If the decision has already been made to mine a pushback and there are no capacity constraints, all blocks that will generate a positive cash flow when processed ($U_{jk,dir} > 0$) should be processed. The decision to process a block is independent of the discount rate. Consequently, under the assumption of no capacity constraint, all blocks that generate a positive cash flow will contribute positively to defining the last pushback and, therefore, the size of the pit. However, optimization of mine and mill operations implies balancing capital and operating costs, which invariably results in capacity constraints and non-zero opportunity costs ($U_{jk,opp} < 0$). Non-zero opportunity costs result in higher cut-off grades, fewer blocks being processed, and, therefore, lower pushback NPV. A pushback that has a positive NPV when capacity constraints are ignored may have a negative NPV if these constraints are taken into account. Ignoring capacity constraints may result in mining pushbacks that should not be mined and designing a pit that is larger than it should be.

As an example, consider a copper mining operation that uses a flotation process and sells concentrate. The mill is capacity constrained. Cut-off grades have been optimized to take into account this capacity constraint. Low-grade material that cannot be processed when mined will be stockpiled. When calculating the NPV of a pushback, one must take into account the following:

- For waste material, the time when it is mined
- For material directly fed to the mill, the time when it is mined and processed
- For material sent to a low-grade stockpile, the time when it is mined as well as the time when it is processed, which is likely to be much later

When optimizing the size of a pushback, one must take into account not only the increase in cut-off grade imposed by capacity constraints but also the time difference between when stockpiled material is mined and when it is processed and copper is sold. If this time difference is ignored, the pushback NPV will be significantly overestimated and low-grade pushbacks may be included in the mine plan that should not be mined.

UNDERGROUND MINE: ECONOMIC VALUATION OF A STOPE

The same situation can occur in underground mines. A stope should be mined if the NPV of generated cash flow is positive. All costs and benefits must be taken into account, as well as when these costs and benefits are realized. This includes the cost of stope development (such as access drifts and crosscuts); the cost of waste mining, stockpiling, and re-handling; the cost of ore mining, stockpiling, re-handling, and processing; and all costs allocated to low-grade stockpiles, if any. Revenues include those incurred from processing ore directly sent to the mill, as well as those realized at a later date from low-grade stockpiles.

If there is no capacity constraint, all material that can generate a positive cash flow if processed when mined will be processed. But project optimization invariably results in capacity constraints, such as those imposed by shaft and drift haulage capacity, ventilation, maximum speed of development, or mining method. These constraints result in non-zero opportunity costs and higher cut-off grades. When capacity constraints are taken into account, the size of some stopes is likely to be reduced, and some stopes will no longer be considered economically minable.

SIMILARITIES BETWEEN OPEN PIT AND UNDERGROUND MINE PLANNING

As shown in the previous discussion, there are many similarities between questions concerning open pit and underground mines, and the approach that must be followed to answer these questions. Here are some of these questions:

- How do capacity constraints influence cut-off grade and cash flow?
- Which cut-off grade should be used to separate waste material, stockpiled material, and material sent to the processing plant?
- Should a pushback be mined in an open pit mine or a stope be mined in an underground mine?
- Should low-grade material at the bottom of a pushback or surrounding a stope be mined or left in the ground?
- If low-grade material must be mined, should it be wasted, stockpiled, or processed?
- How should the time difference between mining, stockpiling, processing, and selling material be taken into account in designing open pit and underground mines?

CHAPTER SEVEN

Cut-off Grade and Mine Planning—Block and Panel Caving

When a block or panel caving mining method is used, estimation of cut-off grades must take into account the limited flexibility that operators have in controlling the grade of material pulled. Cut-off grades are used to determine the location and size of a block or panel, and to decide when pulling material from a draw point should be stopped. Cut-off grades are not likely to play a significant role, if any, when waste or low-grade material is encountered in the middle of a block.

CONSTRAINTS IMPOSED BY BLOCK AND PANEL CAVING

Many factors must be taken into account when designing a block in addition to the geotechnical properties of the deposit and the continuity of mineralization. Ideally, blocks are located in relatively high-grade areas that can be mined without significant internal or external waste dilution, the draw points and production levels are located in lower-grade or waste areas, and the block boundaries are located near lower-grade or waste zones. Internal and external waste or low-grade dilution will occur, which must be taken into account when locating blocks and draw points. When ore is drawn, waste is mixed with higher-grade material, eliminating the opportunity to mine waste selectively.

The rate at which material is pulled from draw points should match the natural rate of caving. The material should be drawn in a uniform fashion across draw points. Production cannot be stopped in one draw point without affecting surrounding draw points. If a draw point containing waste is surrounded by other high-grade draw points, mining waste cannot be stopped. However, if the waste draw point is located on the periphery of the block being mined, this draw point can be stopped. Production is stopped when waste indicates that the entire ore column has been pulled.

Productivity is dependent on a high rate of production, which is not conducive to selective mining of ore and waste material. The capital cost of underground and surface infrastructure needed to handle waste separately from ore grade material is likely to be high. Attempting selective mining is likely to increase mine operating costs significantly. For these reasons, some block caving operations have chosen to send all material mined to the processing plant, whatever the grade.

MARGINAL CUT-OFF GRADE AND DRAW POINT MANAGEMENT

Once a block has been developed and the infrastructure is in place (including drifts, haulage facilities, draw points, ventilation, etc.), the utility of mining and processing one metric ton of material is

$$U_{dir}(x) = x \cdot r \cdot (V - R) - (M + P + O)$$

x = average grade
r = recovery or proportion of valuable product recovered from the mined material
V = value of one unit of valuable product
R = refining, transportation, and other costs that are related to the unit of valuable material produced
M = mining cost per metric ton processed
P = proccessing cost per metric ton processed
O = overhead cost per metric ton processed

The minimum grade that can be mined and processed at a profit is x_{c1} such that $U_{dir}(x_{c1}) = 0$:

$$x_{c1} = [M + P + O]/[r \cdot (V - R)]$$

This cut-off grade should be used to decide whether production from a draw point should be stopped because of excessive lateral dilution or because the entire ore column has been mined.

MARGINAL CUT-OFF GRADE AND BLOCK DESIGN

Incremental analysis must be used to determine the optimal size and location of a block. To decide whether a new row of draw points should be added along the periphery of a block, one must first estimate the tonnage T and average grade x of the material that will be pulled from these draw points, taking dilution into account. If one considers only operating costs and ignores the capital and opportunity cost of adding one row of draw points, this row should be added if the average grade x exceeds the cut-off grade x_{c1} calculated previously.

If the average grade of the last row of draw points is equal to x_{c1}, the cash flow generated from these draw points will not justify the capital cost of developing them. In addition, development of a larger block by addition of peripheral draw points will delay production from what could have been a smaller block. The cut-off grade applicable to the last row of draw points must take into account capital and opportunity costs.

INFLUENCE OF CAPITAL COST AND DISCOUNT RATE

Additional capital expenditures are needed to develop one more row of draw points. This capital cost I must be recovered from profits generated by the draw points. On an undiscounted basis, the profit made from mining and processing T metric tons of material with average grade x is $T \cdot [x \cdot r \cdot (V - R) - (M + P + O)]$. This profit must be greater than or equal to the capital cost I. The cut-off grade applicable to this last row of draw points is determined by adding the capital cost per metric ton I/T to the operating costs M, P and O:

$$x_{c2} = [M + P + O + I/T]/[r \cdot (V - R)]$$

I = capital cost incurred to develop a new row of draw points
T = tonnage to be mined from the new row of draw points

The requirement of a minimum rate of return should be taken into account in calculating the cut-off grade. The following additional notations are used:

i = minimum rate of return (discount rate)
n = number of years during which material will be pulled from the new draw points

Then make the simplifying assumption that the tonnage mined and corresponding average grade will be the same every year, T/n and x, respectively. The yearly cash flow (YCF) expected to be generated from the new draw points is:

$$YCF = (T/n) \cdot [x \cdot r \cdot (V - R) - (M + P + O)]$$

The net present value (NPV) of this cash flow is:

$$NPV = YCF \cdot [1/(1 + i) + 1/(1 + i)^2 + \ldots + 1/(1 + i)^n]$$
$$= YCF \cdot [1 - 1/(1 + i)^n]/i$$

The minimum cut-off grade applicable to the new row of draw points is x_{c3}, such that the net present value of generated cash flows, NPV, is equal to the capital investment, I:

$$NPV = I$$

$$(T/n) \cdot [x_{c3} \cdot r \cdot (V - R) - (M + P + O)] \cdot [1 - 1/(1 + i)^n]/i = I$$

$$x_{c3} = [M + P + O + f(i, n) \cdot (I/T)]/[r \cdot (V - R)]$$

where

$$f(i,n) = n \cdot i/[1 - 1/(1+ i)^n]$$

OPPORTUNITY COST

In addition to increasing capital costs, increasing the size of a block can delay production that could be pulled from a smaller block. Assume that a small, presumably high-grade block has been designed and that a production schedule has been developed accordingly. The net present value NPV_o of future cash flows expected to be generated from mining this block was calculated using the discount rate i. If t is the time by which production from the smaller block will be delayed to allow development of one more row of draw points, the corresponding opportunity cost is

$$U_{opp}(x) = -t \cdot i \cdot NPV_o$$

This opportunity cost represents a decrease in NPV, which must be added to the capital cost of adding the new draw points. Taking this cost into account, the cut-off grade is as follows:

$$x_{c4} = [M + P + O + f(i, n) \cdot (I + t \cdot i \cdot NPV_o)/T]/[r \cdot (V - R)]$$

M = mining cost per metric ton processed
P = processing cost per metric ton processed
O = overhead cost per metric ton processed
$f(i,n) = n \cdot i/[1 - 1/(1 + i)^n]$
n = number of years during which material will be pulled from the new row of draw points
i = minimum rate of return (discount rate)
I = capital cost incurred to develop the new row of draw points
t = time by which previously scheduled production will be delayed
NPV_o = net present value of previously scheduled production

T = tonnage to be mined from the new row of draw points
r = recovery, or proportion of valuable product recovered from the mined material
V = value of one unit of valuable product
R = refining, transportation, and other costs that are related to the unit of valuable material produced

CHAPTER EIGHT

Which Costs Should Be Included in Cut-off Grade Calculations?

Mining engineers face a significant challenge when determining which costs should be included in a cut-off grade calculation. Collaboration between engineers and accountants is necessary to ensure that meaningful numbers are used and that all applicable costs are included. In this chapter, some general principles concerning costs and how they should be treated in the estimation of cut-off grades are discussed.

Costs can be divided between fixed and variable. Fixed costs are expenses for which the total does not change in proportion to the level of activity within the relevant time period or scale of production. By contrast, variable costs change in relation to the level of activity. In cut-off grade calculations, costs incurred when drilling, sampling, blasting, loading, crushing, and grinding the ore; during flotation, concentrate drying, filtering and shipping, smelting and refining, and so forth, are usually considered variable costs. These costs are directly related to the production capacity. Initial capital expenditures, equipment depreciation, general administration, property taxes, marketing, public relations, government relations, and so on, are considered fixed costs. To the extent that fixed and variable costs are properly defined, cut-off grade optimization need only take variable costs into consideration.

However, it is important to realize that fixed costs are fixed only within a certain range of activity or over a certain period of time. If significant changes are made to the cut-off grade that require expansion of a leach pad or tailings dam, costs related to such expansions can no longer be considered fixed. If the life of mine is extended or shortened beyond the current expected life, general and administrative costs will change. These changes should be taken into account in the cut-off grade calculation by allocating their cost to that part of the operation (mine, mill, leach plant, concentrator, smelter, refinery, etc.) that drives the change.

Sunk costs are costs that were incurred in the past and that do not change with the level of activity. Once a mine is in full production, the costs incurred for pre-stripping, shaft sinking, plant construction, and original infrastructure are sunk. Such costs are not taken into account when deciding whether the cut-off grade should be changed. However, during the project feasibility study, all costs, including the initial capital cost, have an influence on the cut-off grade. The cut-off grade determines the tonnage, grade, and location of material available for processing, which in turn drive mine and plant size, capital and operating costs, and financial performance. But operating costs are a critical input in the determination of the minimum cut-off grade. Different cut-off grade profiles, including cut-off grades that decrease over time, will require different mine plans and capital costs and will result in better or worse financial performance. An iterative process must be used to determine the combination of cut-off grades, size of operation, and resulting capital and operating costs that will best satisfy the company's objectives.

Balancing initial and sustaining capital costs, operating costs, and cut-off grades is a critical part of a project feasibility study. If all assumptions made during the feasibility study, including those related to the geology of the deposit, the production capacity, the cost of operations, and the value of the product sold, remained true during the entire life of the mine, the cut-off grades would remain as planned. No cut-off grade change could be justified because plans were optimized and changes would reduce the value of the project. Decreasing the cut-off grade would require that additional lower-grade material be processed, which could not be achieved without either increasing the size of the plant or decreasing the net present value of future cash flows. Conversely, increasing the cut-off grade above that planned would result in underutilization of available capacity.

In practice, operating conditions differ from those assumed during the feasibility study, the geological properties of the deposit differ from those initially expected, capacities are either not reached or exceeded, mine and mill are no longer balanced, costs and the value of products sold are better or worse than expected, and cut-off grades must be continuously re-estimated.

A company's financial objectives are likely to include expectation of a minimum return on investment, which cut-off grade calculations must take into account. If the time needed to mine one metric ton of material, process it, recover a salable product, and get a return from the sale of this product exceeds one year, costs and revenues should be discounted at the company-specified rate. While sunk costs do not influence cut-off grades, the cost of future sustaining capital expenditures must be included in the cut-off grade calculation to ensure that all material processed covers the capital invested, including a specified minimum return on investment.

A few examples follow in which it is assumed that the mining company expects a minimum 15% return ($i = 15\%$) on all investments:

- *Stockpiling of low-grade material.* This was discussed previously. The decision to stockpile material is more often than not a strategic decision rather than a decision based solely on expected cash flows and net present value.
- *Leaching operation.* Consider a leaching operation for which the final recovery is expected to be 80%. This maximum recovery is expected to be reached over three years, being 60% the first year, 12% the second year, and 8% the last year. Revenues from sales will take place over three years and must be discounted to year 1. This can be done by discounting the recovery as follows:

$$\begin{aligned}\text{discounted recovery} &= 60\% + 12\%/(1 + 0.15) + 8\%/(1 + 0.15)^2 \\ &= 76.48\%\end{aligned}$$

  The cut-off grade between wasted and leached material, or between leached and milled material, must be calculated assuming 76.48% leach recovery instead of 80%.
- *Sustaining capital.* Sustaining capital represents capital expenditures that must be incurred on a periodic basis to maintain production at the current level. For example, new trucks may have to be bought every eight years, leach pad expansions may be needed every four years, tails dam lifts may be added every seven years. Let I be the total cost of this investment and n its expected useful life in years. The cut-off grade should be high enough to ensure a minimum return on investment ($i = 15\%$). This is achieved by including the cost of capital in the cut-off grade calculation. Let C_I be the cost per year that must be recognized to recover the investment I over n years at the specified discount rate i. This cost must satisfy the following equation:

$$\begin{aligned}I &= C_I/(1 + i) + C_I/(1 + i)^2 + \ldots + C_I/(1 + i)^{n-1} + C_I/(1 + i)^n \\ &= C_I[1 - 1/(1 + i)^n]/i\end{aligned}$$

  Therefore, the cost per year that should be included in the cut-off grade calculation is

$$C_I = Ii/[1 - 1/(1 + i)^n]$$

$$C_I = (I/n) \cdot f(i, n)$$

$$f(i, n) = n \cdot i/[1 - 1/(1 + i)^n]$$

If i = 15% and n = 8, the cost of capital is $C_I = 0.22$ I per year. If no minimum return on investment was specified, the cost of capital would be $C_I = I/n = 0.13$ I per year.

In the cut-off grade calculation, costs per year must be converted to costs per unit of production. These costs must be added to mining costs if the sustaining capital is for mining equipment, to leaching cost if it is for leach pad expansion, or to milling costs if it is for a tailings dam.

- *Incremental capital expenditures*. Such expenditures may be required to maintain production beyond the planned life, or to reach a higher level of production. The cost of these incremental capital expenditures must be taken into account in the cut-off grade calculation. This is done using the same formula as given previously, where n is the expected useful life of the new infrastructure or equipment.
- *Overhead costs*. General and administration costs (G&A) and other overhead costs must also be divided between fixed and variable costs. Variable G&A costs must be included in all cut-off grade calculations. Fixed G&A costs, usually expressed on a per-year basis, must be included if the change in cut-off grade will change the mine life. This will be the case whenever one of the processes is capacity constrained. The fixed part of overhead costs can no longer be considered as fixed because lowering the cut-off grade will require extending the mine life. These costs must be expressed on a per-unit-of-production basis (by dividing costs per year by production per year) and added to the unit cost of the capacity-constrained process.

CHAPTER NINE

When Marginal Analysis No Longer Applies: A Gold Leaching Operation

In the 1990s, many gold mining companies significantly increased the tonnage of material placed on their leach pads by lowering the cut-off grade. Marginal analysis of leaching costs indicated that already low cut-off grades, often less than 0.5 gram/metric ton, could be further reduced, sometimes down to 0.2 gram/metric ton. The expectation was that, with more ounces being placed on the leach pad, the amount of gold recovered would increase on a monthly basis, as well as cumulatively over time, while the cost per metric ton placed would decrease. The results initially obtained were often disappointing. The tonnage of material added by lowering the cut-off grade was large, resulting in a short-term decrease in recovery, which in the worst cases meant a decrease in revenue instead of the expected increase. In addition, the long-term impact of adding large tonnages of very low-grade material to a leach pad was not fully understood. In some cases, the result seemed to be a decrease in overall pad recovery, not only postponing short-term revenues but showing no increase in cumulative revenues over the life of the project. On a discounted basis, the benefit of lowering the cut-off grade was significantly less than expected, if not negative.

To illustrate this point, consider a gold mining operation where the total tonnage of ore and waste material scheduled to be mined in the coming year was 10 million metric tons. This material was characterized by the grade–tonnage curve shown in Figure 9-1. Initially, the cut-off grade was set at 0.50 gram/metric ton, which corresponded to 6.59 million metric tons of leach-grade material averaging 1.36 grams/metric ton and containing 288,000 ounces of gold. The leach recovery was expected to be 65%, resulting in the production of 187,000 ounces in the coming year.

A review of the previous year's operating costs showed that the cut-off grade could be lowered to 0.40 gram/metric ton if the recovery could be maintained at 65%. Laboratory tests confirmed that recovery was independent of

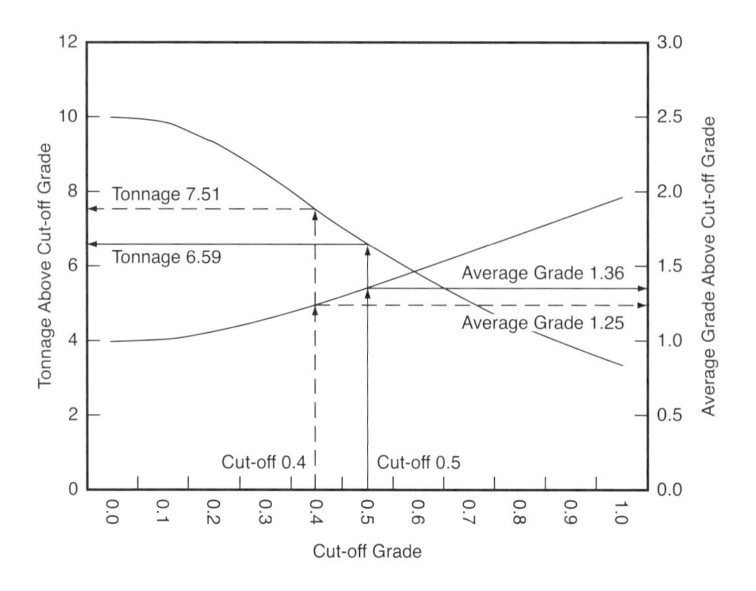

FIGURE 9-1 Estimation of tonnage and grade above cut-off grade

grade and the decision was made to lower the cut-off grade and add the lower-grade material to the pad.

After two months of operation, managers realized that the gold production target for the year was not going to be met. If nothing changed, the amount of gold sold was going to be less than expected before the cut-off grade was decreased. Management immediately requested a review of the situation. The results of this review were as follows:

- Metallurgical tests confirmed no decrease in recovery for lower-grade material.
- Metallurgical tests and review of past operational conditions showed that the amount of gold recovered was an increasing function of the solution ratio, defined as the metric tons of cyanide solution used per metric ton of material placed on the pad. This relationship was as illustrated in Figure 9-2.
- Provided that a four-month leaching cycle was adhered to, the original solution ratio was 1:1, as needed to reach 65% recovery.
- Lowering the cut-off grade to 0.40 gram/metric ton increased the tonnage to be placed on the pad from 6.59 to 7.51 million metric tons and decreased the average grade from 1.36 to 1.25 grams/metric ton (Figure 9-1). The ounces placed increased from 288,000 ounces to 302,000 ounces, a 5% increase.

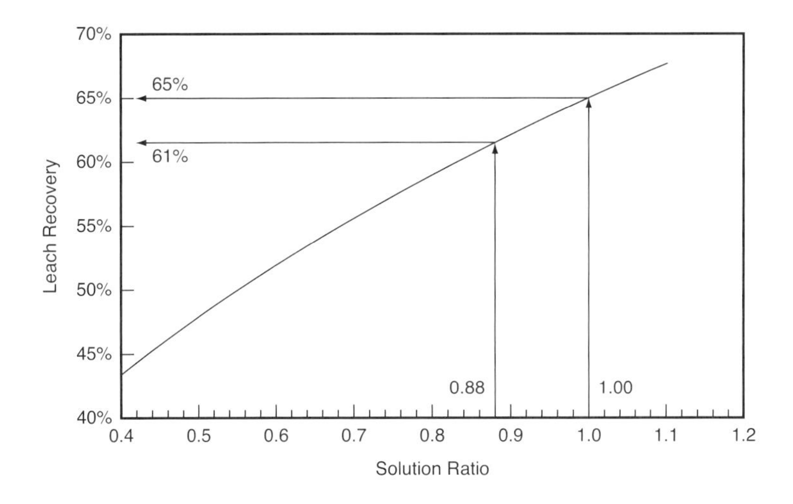

FIGURE 9-2 Relationship between leach recovery and solution ratio

- Because no change was made to the amount of solution placed on the pad, the increase in tonnage from 6.59 to 7.51 million metric tons resulted in a decrease in solution ratio from 1.0 to 0.88. The expected recovery should have been 61% instead of 65% (Figure 9-2).
- This 6% decrease in recovery exceeds the expected 5% increase in ounces placed on the pad. The total metal recovered during the year should have been expected to decrease from 187,000 ounces to 184,000 ounces.

Ignoring the relationship between leach recovery and solution ratio was equivalent to ignoring a capacity constraint. The corresponding opportunity cost was ignored, and consequently the cut-off grade was underestimated. Lowering the cut-off grade to 0.40 gram/metric ton might have been justified if a cost-effective method of increasing the recovery had been put in place. One option was to increase the volume of fresh solution placed on the pad, which would require changes in pipes, pumps, and the capacity of the carbon columns or Merrill–Crowe plant used to process the solution. Another option was to recycle the pregnant solution on the pad, which would increase the solution-to-ore ratio without incurring some of the high costs associated with the first option. All changes to the leach plant had to take into account constraints imposed by operating permits, pond size, and other conditions (technical, environmental, or legal), which would limit the options available to solve the problem.

Assume that, for environmental and permitting reasons, the size of the leach plant could not be increased. Which approach should have been used to determine the optimal cut-off grade? Taking into account the low operating costs, this optimal grade is likely to be less than 0.5 gram/metric ton (which was determined on the basis of higher costs) but more than 0.4 gram/metric ton (which used the lower costs but ignored the operating constraint). An iterative approach could be used that consists of decreasing the cut-off grade by small successive increments and fully assessing the economic consequences until no further decrease is justified.

1. Assume that the cut-off grade is lowered from the current 0.50 gram/metric ton to 0.48 gram/metric ton.
2. Estimate the increase in tonnage and ounces that will be placed on the pad as a result of the lower cut-off grade.
3. Calculate the corresponding decrease in solution ratio and leach recovery.
4. Calculate the resulting change in total gold recovered, taking into account the increase in gold placed and decrease in recovery.
5. Compare the change in expected gold sold with the corresponding change in cost of operation. Differences between the cost of wasting material and placing it on the pad should be taken into account.
6. If the change in revenue from sales exceeds the change in costs, the cut-off grade can be reduced to 0.48 gram/metric ton. The analysis should then be repeated, assuming a lower 0.46 gram/metric ton cut-off. The optimum cut-off is that for which the change in revenue is equal to the change in cost.

CHAPTER TEN

Mining Capacity and Cut-off Grade When Processing Capacity Is Fixed

Ideally, a new mine should be designed such that mining capacity and processing capacity are perfectly balanced and the planned cut-off grades fill up these capacities and result in optimized expected cash flow. In practice, this situation occurs only on paper, when the project is designed. As soon as operations start, imbalances invariably appear. The actual processing plant capacity exceeds or falls below that expected. The mining capacity is higher or lower than planned. Mine and mill capacities are no longer balanced, new constraints appear, and the cut-off grade must be changed accordingly. The cut-off grade must also take into account differences between expected and actual costs, productivities, recoveries, and market value of product sold. When a new project is designed, mine and mill capacities and corresponding cut-off grades are chosen primarily to optimize financial objectives. Once mine and mill facilities are built, physical constraints become the main drivers and studies must be completed to determine whether removing these constraints is financially justified.

In this chapter, it will be assumed that the capacity of the processing plant is fixed and cannot be changed. The only change that can be made is to the mining capacity. What is the impact of a change in mining capacity on the cut-off grade and the grade of the material sent to the plant?

- Consider an increase in mining capacity, defined as tonnage mined per year.
- This increase requires an increase in mining capital cost, and is likely to result in an increase in total mine operating costs per year. However, it is also likely to result in decreased mining and overhead costs per metric ton mined.
- The processing capacity being fixed, the cut-off grade must be increased to keep the tonnage sent to the mill constant. The average grade of mill feed will increase and so will the quantity of product sold.

- The mine life will decrease.
- In some instances, the lower-grade material that is not processed will be stockpiled. Stockpiling of low-grade material was discussed previously.

To decide whether an increase in mining capacity is justified, the expected net impact on the utility of the project must be assessed, taking into account a variety of factors:

- Increased capital cost of new mining capacity
- Decreased mine unit operating cost
- Increased plant head grade and increased metal sales per year
- Loss of low-grade material or delayed processing of some of this material
- Reduced mine life and resulting socio-economic and political impact
- Reduced project life and decreased political risk if applicable
- Change in environmental impact

A simple example follows. Consider a mining operation in which the plant was designed to process an average of 250,000 metric tons per month, or 3 million metric tons per year. The grade–tonnage relationship corresponding to the mineralized material expected to be mined during the coming year is shown in Figure 10-1. At the current mining capacity, the 3 million metric ton plant capacity is consistent with a cut-off grade of 0.74 gram/metric ton and a mill feed average grade of 1.56 grams/metric ton.

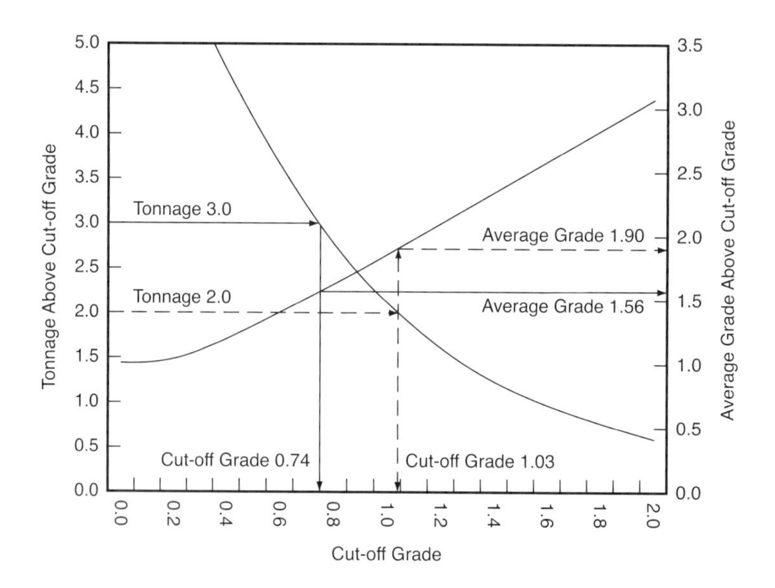

FIGURE 10-1 Estimation of cut-off grade assuming fixed processing capacity

Management is considering increasing the mining capacity by 50% and is investigating the impact such a change would have on the coming year. If the mining capacity is increased by 50%, the same material shown in Figure 5.1 will be mined in eight months instead of one year. Because the mill can process only 250,000 metric tons per month, it will only consume 2 million metric tons during the eight-month period. This tonnage corresponds to a cut-off grade of 1.03 grams/metric ton and an average mill feed head grade of 1.90 grams/metric ton. Consideration should be given to stockpiling the material between 1.03 grams/metric ton and a cut-off grade somewhat higher than 0.74 gram/metric ton. This material should be considered for re-handling and processing at a later date.

The proposed 50% increase in mining capacity may or may not be optimal. To be economically justified, an increase in mining capacity must take into account financial, technical, environmental, permitting, and other constraints imposed by deposit size and shape, mining method, size of equipment, safety and environmental regulations, and other parameters. Depending on the limitations imposed by these constraints, an iterative approach is best suited to mining capacity optimization. Such an approach can consist of the following steps:

1. Assume a 1-million-metric-ton increase in mining capacity (or some other increase that is technically achievable).
2. Calculate the resulting decrease in mine life.
3. Estimate the increase in cut-off grade and resulting higher mill head grade that is consistent with the increase in mining capacity and fixed processing capacity.
4. Estimate the increase in mine capital and yearly operating costs needed to increase the mining capacity. Calculate the corresponding discounted incremental mining cost (DIMC) for the remaining life of the project.
5. Estimate the increase in mill production per year (units of product sold) and calculate the corresponding discounted incremental revenue (DIR).
6. If low-grade material is to be stockpiled, the net present value of this material should also be taken into account.
7. If the DIR exceeds the DIMC, this analysis should be repeated, assuming an additional 1-million-metric-ton increase in mining capacity.
8. The optimal mining capacity is that for which DIR is equal to DIMC.

In the previous discussion, it was assumed that the increase in mine capacity could be achieved without changing mine selectivity. The grade–tonnage curve did not change. The volumes being mined remained the same but these volumes were mined faster. This situation will occur if more equipment of the same size is added to an open pit mine with no change to the pit design or if more stopes are put in production simultaneously without changing the underground mining method or the stope design. There are situations where the assumptions of constant grade–tonnage curve cannot be made. In open pit mines, capacity can be increased by using larger trucks and loading equipment, increasing the bench height, and widening the spacing between blast holes. The result is a decrease in selectivity, resulting in a new grade–tonnage curve. Similarly, underground production can be increased by using a different mining method that will be less selective but results in significantly lower costs per metric ton. The impact of selectivity on the grade–tonnage curve, the cut-off grade, and the mill feed average grade will be discussed later. Increasing the mining capacity will not necessarily result in a higher head grade if this increase is realized by significantly decreasing mine selectivity.

CHAPTER ELEVEN

Processing Capacity and Cut-off Grade When Mining Capacity Is Fixed

In the previous chapter, a fixed processing capacity was assumed to be the case. Now consider the case where the mining capacity is fixed, but an increase in plant capacity is being contemplated. A lower cut-off grade is needed to balance the mining capacity with the plant capacity. Increasing the mill capacity has the following impacts:

- The tonnage processed per year is increased.
- The tonnage mined is not changed. The cut-off grade must be decreased to keep the processing plant full.
- The average grade of material sent to the mill decreases, but the metal content of this material increases.
- More metal is recovered, resulting in higher revenues from sales.
- The capital cost of plant expansion must be taken into account.
- The plant operating costs are likely to increase per unit of time (cost per year) but to decrease per unit of production (cost per metric ton processed).

The optimal plant capacity is that which maximizes the total utility of the project, taking into account financial impact (increased capital cost, decreased unit operating cost, increased revenue from sales), as well as socio-economic, environmental, political, and other impacts.

A simple example follows. Consider a mining operation in which the plant was designed to process an average of 250,000 metric tons per month, or 3 million metric tons per year. The grade–tonnage relationship corresponding to the mineralized material expected to be mined at the current capacity is shown in Figure 11-1. At the current mining capacity, this plant capacity is consistent with a cut-off grade of 0.74 gram/metric ton, corresponding to a mill feed average grade of 1.56 grams/metric ton.

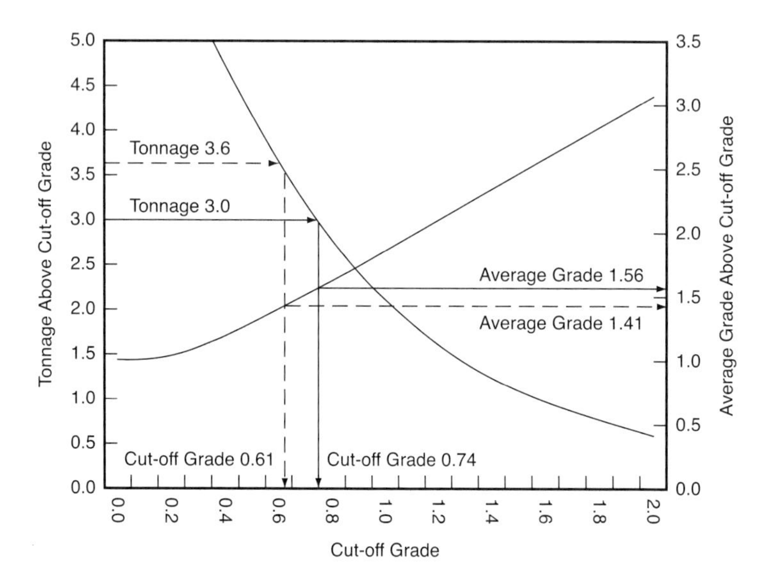

FIGURE 11-1 Estimation of cut-off grade assuming fixed mining capacity

Management is considering increasing the size of the processing plant by 20% and is investigating the impact that such a change would have on the coming year. If the processing capacity is increased by 20% and the mining capacity is kept constant, the cut-off grade must be decreased to 0.61 gram/metric ton to supply 3.6 million metric tons to the mill (Figure 11-1), and the average grade of mill feed will decrease to 1.41 grams/metric ton. The gold content of the material processed will increase from 2.51 million ounces to 2.73 million ounces. This cut-off grade calculation only takes into account capacity constraints and is independent of the economics of the project. The increase in plant capacity must be justified not only by the increase in material processed, but also by taking into account capital cost requirements, possible changes (increase or decrease) in recovery, likely decrease in operating costs, and all other direct and indirect costs and benefits.

The proposed 20% increase in processing capacity may or may not be optimal. To be economically justified, an increase in plant capacity must take into account financial, technical, environmental, permitting, and other constraints imposed by the size of the available processing equipment, limitations on tailings dam expansion, maximum permitted dust emission, and other parameters. Depending on the limitations imposed by these constraints, an iterative approach is best suited to plant capacity optimization. This approach can consist of the following steps:

1. Assume a 1-million-metric-ton increase in processing capacity (or some other increase that is technically achievable).
2. Estimate the decrease in cut-off grade and resulting lower mill head grade that is consistent with the fixed mining capacity and higher processing capacity.
3. Estimate the increase in mill capital and yearly operating costs needed to increase the processing capacity. Calculate the corresponding discounted incremental processing cost (DIPC) for the remaining life of the project.
4. Estimate the increase in mill production per year (units of product sold) and calculate the corresponding discounted incremental revenue (DIR).
5. If the DIR exceeds the DIPC, this analysis should be repeated, assuming an additional 1-million-metric-ton increase in processing capacity.
6. The optimal processing capacity is that for which DIR is equal to DIPC.

CHAPTER TWELVE

Mining and Processing Capacity and Cut-off Grade When Sales Volume Is Fixed

In this chapter, it is assumed that the volume of sales is fixed. This may be because all products are sold under contracts that specify the volume that will be bought on a yearly basis. Perhaps the market is small and the amount of product that can be sold is limited. Or it might be because management specifies the amount to be produced from a given operation for reasons external to the operation under consideration.

FIXED SALES WITH NO MINING OR PROCESSING CONSTRAINT

Provided the recovery achieved in the processing plant is independent of tonnage processed and plant head grade, assuming a fixed volume of sales is equivalent to assuming a fixed quantity of metal (or other salable product) delivered by the mine to the processing plant. This quantity Q_{+c} is equal to the tonnage delivered T_{+c} multiplied by the average grade of plant feed x_{+c}:

$$Q_{+c} = T_{+c} \cdot x_{+c}$$

Consider a gold mining operation that has been requested to supply four metric tons of gold (130,000 ounces) to the processing plant over a one-year period (Q_{+c} = 4.0 metric tons of gold). Consider three scenarios:

- There is no constraint on either mine or plant capacity. This is usually only the case during the feasibility study.
- The mine capacity is fixed, but the plant capacity is not.
- The plant capacity is fixed, but the mine capacity is not.

If neither the mine nor the processing plant is capacity constrained, the number of possible cut-off grades is theoretically infinite. A high cut-off grade will result in a high average grade above cut-off grade x_{+c}. The higher the cut-off grade, the lower the capacity T_{+c} of the processing plant that is needed to keep sales at the required level. But in the case of an open pit mine, a higher cut-off grade will require mining more metric tons per year. In the case of an underground mine, smaller stopes may have to be designed to eliminate peripheral low-grade material, and low-grade stopes may have to be rejected.

When neither mine nor plant capacity is fixed, cut-off grade optimization requires analysis of a number of feasible solutions: low cut-off grade and large plant size, or high cut-off grade and smaller plant size. Technical constraints, including constraints imposed by the geology of the deposit, will reduce the number of feasible options. Higher cut-off grades will result in lower capital costs for the plant and likely higher operating costs, while the impact on mine capital and operating costs will be a function of the geological properties of the deposit and the applicable mining methods. Cut-off grade optimization requires estimation of capital and operating costs and cash flow analysis for each feasible solution.

FIXED SALES AND FIXED PROCESSING RATE WITH NO MINING CONSTRAINT

Cut-off grade determination becomes easier if, in addition to the constraint on the amount of metal processed, one adds a constraint on either plant or mine capacity. First assume that the plant capacity, defined as tonnage processed per year, is fixed. With both tonnage processed T_{+c} and metal content Q_{+c} being fixed, the plant head grade x_{+c} is calculated as follows:

$$x_{+c} = Q_{+c}/T_{+c}$$

If one knows what material can be mined in the coming months, one can determine the cut-off grade needed to reach the necessary average grade and the mining rate needed to reach the necessary tonnage of mill feed T_{+c}.

As an example, again consider the gold mine that was asked to supply four metric tons of gold to the processing plant during the coming year. In addition, assume that the capacity of the processing plant is fixed at 2 million metric tons per year. To satisfy these constraints, the head grade must be

$$\begin{aligned} x_{+c} &= Q_{+c}/T_{+c} \\ &= (4{,}000{,}000 \text{ grams/year})/(2{,}000{,}000 \text{ metric tons/year}) \\ &= 2.00 \text{ grams/metric ton} \end{aligned}$$

A preliminary mine plan was developed during which 6 million metric tons of material, both ore and waste, would be mined. The grade–tonnage relationship corresponding to this material is shown in Figure 12-1. From this relationship, one determines that to get an average grade of 2.0 grams/metric

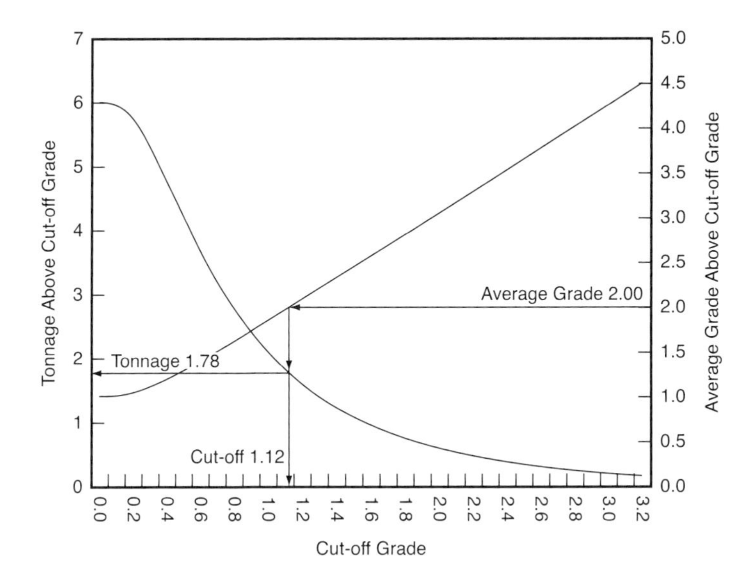

FIGURE 12-1 Estimation of cut-off grade and tonnage given an average grade

ton, one needs to use a cut-off grade of 1.12 grams/metric ton. There are only 1.78 million metric tons of mill feed above cut-off grade in this preliminary mine plan. Because the mill capacity is 2 million metric tons per year, this material will be processed in 10.7 months, calculated as follows:

$$(12 \text{ months/year}) \cdot \frac{(1.78 \text{ million metric tons})}{(2.0 \text{ million metric tons/year})} = 10.7 \text{ months}$$

Six million metric tons are scheduled to be mined in this preliminary mine plan. To mine this tonnage in 10.7 months, the mining rate must be 6.0/10.7 = 560,000 metric tons per month or 6.7 million metric tons per year.

In conclusion, for the mine to send four metric tons of gold per year to a plant that has a capacity of 2 million metric tons per year, a total of 6.7 million metric tons must be mined every year and a cut-off grade of 1.12 grams/metric ton must be used. The plant head grade will be 2.0 grams/metric ton.

FIXED SALES AND FIXED MINING RATE WITH NO PROCESSING CONSTRAINT

Now consider the case where the mine capacity is constrained at 6 million metric tons per year and the metal content of the material to be sent to the mill is set at four metric tons of gold per year. A yearly mine plan was developed in

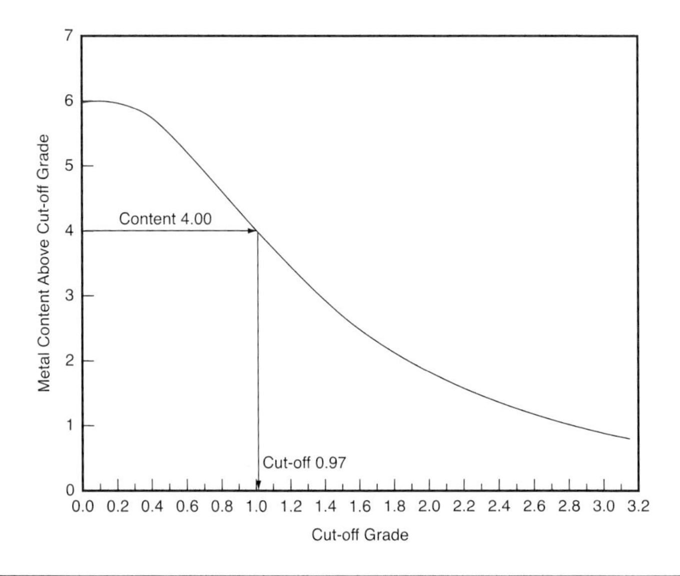

FIGURE 12-2 Estimation of cut-off grade given the required metal content of mine feed

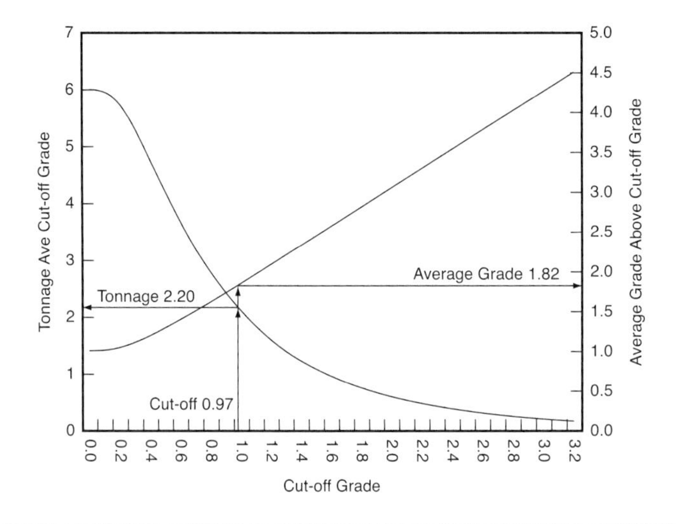

FIGURE 12-3 Estimation of tonnage and average grade above cut-off grade

which 6 million metric tons are to be mined. The corresponding grade–tonnage relationship is shown in Figure 12-1. From the values of T_{+c} and x_{+c} shown in Figure 12-1, it is possible to calculate the metal content of material above cut-off grade $Q_{+c} = T_{+c} \cdot x_{+c}$ and plot this metal content as a function of the cut-off grade x_c (Figure 12-2).

Figure 12-2 shows the relationship between cut-off grade and quantity of metal above cut-off grade, as scheduled to be mined in the current mine plan. Because the quantity of metal to be processed is $Q_{+c} = 4.0$ metric tons of gold, the cut-off grade must be 0.97 gram/metric ton. The tonnage and average grade of material above this cut-off grade can be determined using the grade–tonnage relationship (Figure 12-3):

$$T_{+c} = 2.20 \text{ million metric tons}$$

$$x_{+c} = 1.82 \text{ grams/metric ton}$$

Given that 6 million metric tons of material are scheduled to be mined in the coming year and that the mine must send four metric tons of gold to the processing plant, a cut-off grade of 0.97 gram/metric ton must be used, resulting in 2.20 million metric tons of material being sent to the processing plant, averaging 1.82 grams/metric ton. This can be achieved only if the plant capacity is at least 2.20 million metric tons per year.

CHAPTER THIRTEEN

Releasing Capacity Constraints: A Base Metal Example

In this chapter, a copper mining and processing operation is considered. Mine and mill capacities are 79 million metric tons and 39.5 million metric tons per year, respectively. The copper resources included in that part of the deposit scheduled to be mined in the coming year are listed in Table 13.1. The cut-off grade for mill feed is 0.25%Cu. The reserves to be mined in the coming year are 39.5 million metric tons of ore averaging 0.381%Cu and containing 150,000 metric tons of copper (332 million pounds of copper).

Management wishes to assess the sensitivity of the project to changes in mine, mill, or smelter capacity under a number of conditions. Four cases will

TABLE 13-1 Copper resources contained in material scheduled to be mined

Cut-off, %Cu	Minable Tonnage, million metric tons	Minable Grade, %Cu	Minable Copper Content	
			thousand metric tons Cu	million pounds Cu
0.15	53.7	0.335	180	397
0.16	52.6	0.340	179	395
0.17	51.4	0.344	177	390
0.18	50.1	0.348	174	384
0.19	48.8	0.352	172	378
0.20	47.5	0.355	168	372
0.21	46.0	0.360	165	365
0.22	44.0	0.365	162	357
0.23	42.8	0.370	159	349
0.24	41.2	0.375	155	341
0.25	39.5	0.381	150	332
0.26	37.7	0.387	146	322
0.27	35.9	0.393	141	311
0.28	34.1	0.399	136	300
0.29	32.1	0.406	131	288
0.30	30.2	0.413	125	275
0.31	28.2	0.421	119	262

TABLE 13-2 Cut-off grades, mine and mill capacities required to satisfy specific capacity requirements

	Cut-off Grade, %Cu	Tonnage Milled, million metric tons	Average Grade, %Cu	Copper Content		Tonnage Mined, million metric tons
				thousand metric tons Cu	million pounds Cu	
Base Case						
Value	0.250%	39.5	0.381%	150	332	79.0
Case 1: Increase mining rate by 10%. Keep processing rate at same level.						
Value	0.270%	39.5	0.393%	155	342	86.9
Difference from base case	8%	0%	3%	3%	3%	10%
Case 2: Increase processing rate by 10%. Keep mining rate at same level.						
Value	0.225%	43.5	0.367%	160	352	79.0
Difference from base case	–10%	10%	–4%	6%	6%	0%
Case 3: Increase copper production by 10%. Keep mining rate at same level.						
Value	0.210%	46.0	0.360%	165	365	79.0
Difference from base case	–16%	16%	–6%	10%	10%	0%
Case 4: Increase copper production by 10%. Keep mining rate at same level.						
Value	0.305%	39.5	0.418%	165	364	107
Difference from base case	22%	0%	10%	10%	10%	36%

be considered, which are summarized in Table 13.2. Each case is compared with the base case, in which 79 million metric tons are mined, of which 39.5 million metric tons are processed.

A description of each case follows.

- Case 1: Assume that the mine capacity is increased by 10%, from 79 to 86.9 million metric tons, but the mill capacity remains fixed at 39.5 million metric tons per year. The 79 million metric tons that were scheduled to be mined in one year, including the resources shown in Table 13.1, will be mined in 0.91 years (10.9 months). During this period, the mill can only process 35.9 million metric tons. From Table 13.1 one sees that to send only 35.9 million metric tons to the processing plant, one must increase the cut-off grade to 0.270%Cu. The mill head grade will be 0.393%Cu. Assuming that the same average grade can be maintained over one year, 39.5 million metric tons of ore will be processed at an average grade of 0.393%Cu, containing 155,000 metric tons of copper.

- Case 2: Assume that the capacity of the flotation plant is increased by 10%, from 39.5 to 43.5 million metric tons, but the mine capacity is unchanged at 79 million metric tons per year. The resources available to feed the mill remain as shown in Table 13.1. To supply 43.5 million metric tons to the mill, the cut-off grade must be lowered to 0.225%Cu. The mill head grade will average 0.367%Cu, resulting in 160,000 metric tons of copper being processed.
- Case 3: Management wishes to determine under which conditions 10% more copper could be sent to the processing plant if mine capacity remains fixed at 79 million metric tons. The copper content of processed material must increase from 150,000 metric tons to 165,000 metric tons. From Table 13.1 it can be seen that the cut-off grade must be decreased to 0.21%Cu, resulting in 46.0 million metric tons of ore being sent to the mill averaging 0.360%Cu. If the mining rate is not changed, a 10% increase in copper processed can only be achieved by decreasing the average grade by 6% and increasing the tonnage milled by 16%.
- Case 4: Management wishes to determine under which conditions 10% more copper could be sent to the processing plant if mill capacity remains fixed at 39.5 million metric tons. To increase the copper content of mill feed from 150,000 metric tons to 165,000 metric tons, the mill head grade must be increased from 0.381%Cu to 165,000/39,500,000 = 0.418%Cu. Table 13.1 shows that, to reach this average grade, it is necessary to use a cut-off grade of 0.305%Cu. There are only 29.2 million metric tons above this cut-off grade. Given the mill's capacity of 39.5 million metric tons, this ore will be consumed in 8.86 months. The mining rate must therefore be increased from 79 million metric tons per year to 79 · 12/8.86 = 107 million metric tons per year. If the processing rate is not changed, a 10% increase in copper processed can only be reached by increasing the average grade by 10% and increasing the tonnage mined by 36%.

These examples show procedures that can be used to calculate cut-off grades, taking into account geologic constraints (as summarized in Table 13.1) and technical constraints, including mine, mill, or production limitations. No attempt was made to assess whether the proposed solutions were economically feasible or justified. Implementing any of the mining and processing plans summarized in Table 13.2 would require additional capital expenditures, change operating costs, result in shorter mine life (cases 1 and 4), justify stockpiling of low-grade material (case 4), and require other operational changes, all of which would result in changes in cash flow.

CHAPTER FOURTEEN

Relationship Between Mine Selectivity, Deposit Modeling, Ore Control, and Cut-off Grade

In the previous examples, it was assumed that the grade–tonnage relationship that characterizes the deposit is independent of the mining capacity. However, in many instances, changes in mining capacity are accompanied by changes in mining method, size of mining equipment, bench height, stope dimensions, drill hole spacing, ore control method, and other parameters that determine mine selectivity and the shape of the grade–tonnage curve. These changes must be taken into account in establishing the likely effect that changes in mining capacity and cut-off grade will have on mill feed and reserves.

As an example, consider a deposit for which the total resources above a zero cut-off grade are estimated at 20 million metric tons averaging 10 grams/metric ton. The geology of the deposit is such that either open pit or underground mining methods can be used. Figures 14.1 and 14.2 both show the grade–tonnage relationships corresponding to the open pit and underground mining methods. On Figure 14-1, the resources that can be mined from the deposit using the low-selectivity open pit mining method are shown as solid lines. The resources that can be mined from the same deposit using the high-selectivity underground mining method are shown as dotted lines. On Figure 14-2, the underground resources are shown as solid lines while the open pit resources are shown as dotted lines.

The open pit cut-off grade was estimated at 3.0 grams/metric ton. The amount of material that could be mined above this cut-off grade was 15.2 million metric tons, averaging 12.6 grams/metric ton and containing 6.1 million ounces (solid lines on Figure 14-1). If the high-selectivity model had been used to evaluate the open pit option, the reserves would have been erroneously estimated at 8.6 million metric tons, averaging 21.8 grams/metric ton and containing 6.0 million ounces (dotted lines on Figure 14-1).

The underground cut-off grade was estimated at 10.0 grams/metric ton. The amount of material that could be mined above this cut-off grade was

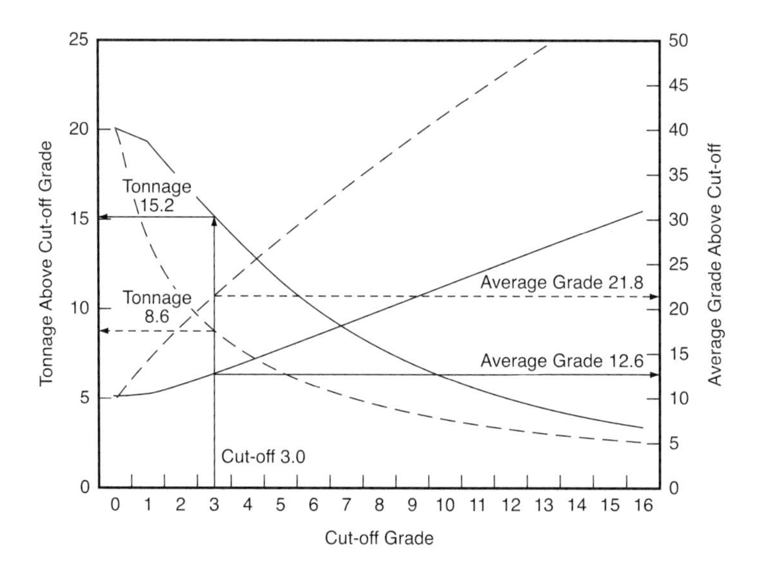

FIGURE 14-1 Application of low-selectivity cut-off grade to low- and high-selectivity models

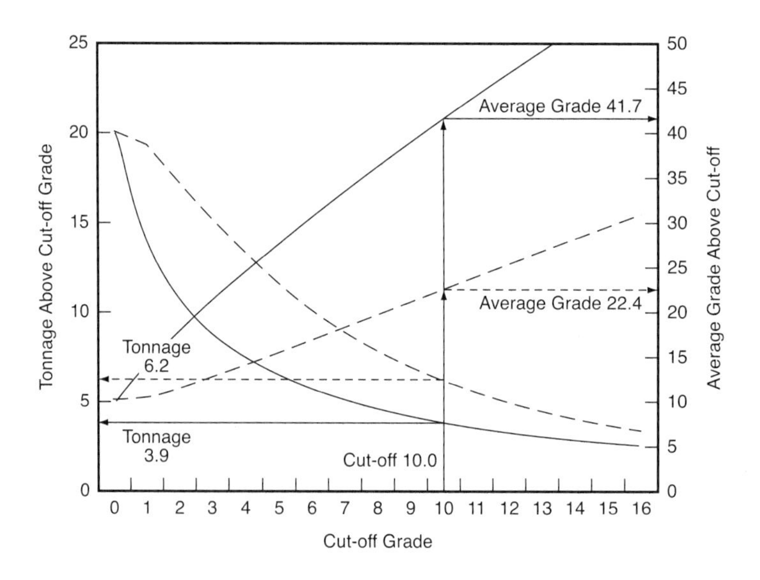

FIGURE 14-2 Application of high-selectivity cut-off grade to high- and low-selectivity models

TABLE 14-1 Influence of deposit model and cut-off grade on mineral reserves

	Underground Mine (Cut-off: 10.0 grams/metric ton)			Open Pit Mine (Cut-off: 3.0 grams/metric ton)		
Deposit Model	Metric tons, millions	Grade, g/metric ton	Ounces, millions	Metric tons, millions	Grade, g/metric ton	Ounces, millions
High selectivity	3.9	41.7	5.2	8.6	21.8	6.0
Low selectivity	6.2	22.4	4.4	15.2	12.6	6.1
Correct model	3.9	41.7	5.2	15.2	12.6	6.1
Error if correct model is not used	60%	–46%	–14%	–43%	73%	–2%

3.9 million metric tons, averaging 41.7 grams/metric ton and containing 5.2 million ounces (solid lines on Figure 14-2). If the low-selectivity model had been used to determine the feasibility of the underground mining method, the reserves would have been erroneously estimated at 6.2 million metric tons, averaging 22.4 grams/metric ton and containing 4.4 million ounces (dotted lines on Figure 14-2).

The errors made when using the open pit model to evaluate the underground resources or the underground model to evaluate the open pit resources are summarized in Table 14-1. While this table represents an extreme case, it clearly shows that changes in mining method and changes in cut-off grade must be evaluated jointly, and that appropriate deposit models must be used which reflect the conditions that are expected to prevail when these changes are made. When assessing the impact that changes in mining capacity may have on mill head grades, one must take into account not only changes in cut-off grades but also changes in the grade–tonnage curve. The grade–tonnage curve will remain the same only if no change is made to mining method, ore control practices, and size of mining equipment.

A computer-generated deposit model is the foundation on which mine plans are developed, cut-off grades are optimized, and the tonnage and average grade of material processed are determined. For the results of a feasibility study to be meaningful, the deposit model must reflect the geological properties of the deposit. In addition, the relationship between cut-off grade, tonnage, and average grade above cut-off grade, which is implied by the deposit model, must be the same as that which will be realized when the deposit is mined.

The deposit model must be developed taking into account the mining method that will be used and how selective this method will be. Different models are usually needed for open pit and underground mines, for bulk mining and selective mining, for block caving and cut-and-fill. Selectivity is a function not only of the geology of the deposit and the mining method but also of bench height and blast hole spacing, stope design, type and size of mining equipment, and ore control method. The significance of these factors must be assessed when developing the deposit model.

It is not sufficient to make realistic selectivity assumptions when developing the deposit model and optimizing the cut-off grade. These assumptions must be respected when the deposit is being mined. Otherwise, the tonnage and average grade of material mined and processed will differ from that estimated when the project feasibility study was completed. In practice, changes will occur during the life of the mine, which will change selectivity. Such changes may include changing mining method, using smaller or higher bench heights, designing larger or smaller stopes, changing the equipment size, and modifying ore control practices. Whenever such changes are made, one must question whether they will change the grade–tonnage curve sufficiently to require development of a new deposit model.

CHAPTER FIFTEEN

Conclusions

The cut-off grade determines the tonnage and average grade of material processed and is critical to determining the economic feasibility of a project. All consequences of choosing a cut-off grade must be taken into account, including technical, economic, legal, environmental, social, and political, as illustrated by the following fundamental equation:

$$U(x) = U_{dir}(x) + U_{opp}(x) + U_{oth}(x)$$

Cut-off grade optimization is an iterative process. When planning a mining operation, a cut-off grade profile must be chosen to define the size of the mine, the capacity of the processing plant, and the resulting cash flow. But the optimal cut-off grade is a function of the cash flow generated by the project. Once the cash flow has been determined, the cut-off must be re-estimated. Cut-off grades must also be revised as planning progresses, when the geology of the deposit is better understood, when the deposit model is updated, when mining and processing methods are better defined, when constraints on production are quantified, and when the achievable mine selectivity is established.

Once a mine is in production, management's expectation is that the cash flow will be similar to that indicated by the feasibility study. However, operational conditions are rarely identical to those assumed during the feasibility study. There are differences between the deposit model developed from exploration data and the actual geological, geotechnical, and metallurgical properties of the deposit. Mine production is either higher or lower than planned. The mill can process more or fewer metric tons than anticipated. The mill recovery is higher or lower than was estimated from metallurgical tests. Capital and operating costs differ from those included in the feasibility study. The price of product sold is not as forecasted. Cut-off grades must be periodically reviewed and changed as operating conditions change. The method used to optimize cut-off grades should be the same throughout the project life, during the feasibility study as well as when the mine is in production. However, the optimal cut-off grade will change as the controlling variables change over time.

Bibliography

Carter, P.G., D.H. Lee, and H. Baarsma. "Optimisation methods for the selection of an underground mining method." In *Proceedings of Orebody Modelling and Strategic Mine Planning Symposium*, ed. R. Dimitrakopoulos and S. Ramazan. Perth, Australia: The Australasian Institute of Mining and Metallurgy, 2004.

Chanda, E.K. "Network linear programming optimization of an integrated mining and metallurgical complex." In *Proceedings of Orebody Modelling and Strategic Mine Planning Symposium*, ed. R. Dimitrakopoulos and S. Ramazan. Perth, Australia: The Australasian Institute of Mining and Metallurgy, 2004.

Dagdelen, K., and K. Kawahata. "Value creation through strategic mine planning and cutoff-grade optimization," *Mining Engineering* 60 (2008): 39–45.

Dimitrakopoulos, R., L. Martinez, and S. Ramazan. "Optimizing open pit design with simulated orebodies and Whittle Four-X: A maximum upside/minimum downside approach." In *Proceedings of Orebody Modeling and Strategic Mine Planning Symposium*, ed. R. Dimitrakopoulos and S. Ramazan. Perth, Australia: The Australasian Institute of Mining and Metallurgy, 2004.

Hall, B.E. "How mining companies improve share price by destroying shareholder value." Paper 1194 in *Proceedings CIM Mining Conference and Exhibition*, Montreal: Canadian Institute of Mining and Metallurgy, 2003.

Hoerger, S., J. Bachmann, K. Criss, and E. Shortridge. "Longterm mine and process scheduling at Newmont's Nevada operations." In *Proceedings of 28th APCOM Symposium*, ed. K. Dagdelen. Golden, Colorado: Colorado School of Mines, 1999.

Hoerger, S., Hoffman, L., and F. Seymour. "Mine planning at Newmont's Nevada operations," *Mining Engineering* 51 (1999): 26–30.

King, B. "Optimal mine scheduling." In *Monograph 23: Mineral resource and ore reserve estimation—the AusIMM guide to good practice*, ed. A.C. Edwards. Carlton, Victoria, Australia: The Australasian Institute of Mining and Metallurgy, 2001.

Lane, K.F. *The economic definition of ore: cut off grades in theory and practice*. London: Mining Journal Books, 1991. First published 1988.

Lerchs, H., and I.F. Grossmann. "Optimum design of open-pit mines." *CIM Bulletin 58*, no. 633 (1965): 47–54.

Whittle, J. "A decade of open pit mine planning and optimization—The craft of turning algorithms into packages." In *Proceedings of 28th APCOM Symposium*, ed. K. Dagdelen. Golden, Colorado: Colorado School of Mines, 1999.

Symbols

Symbol	Description
c	constant tail
C_I	revenues required every year during n years to get a return on investment i on a capital investment I: $C_I = Ii/[1 - 1/(1 + i)^n]$
C_s	smelter costs per metric ton of concentrate
C_t	cost of shipping one metric ton of concentrate to the smelter
d_1	metal grade deducted from recovered grade in calculation of smelter payment for metal 1
d_2	metal grade deducted from recovered grade in calculation of smelter payment for metal 2
DIMC	discounted incremental mining cost
DIPC	discounted incremental processing cost
DIR	discounted incremental revenue
$dP_o(T_{+c})dT_{+c}$	First-order derivative of $P_o(T_{+c})$ with respect to T_{+c}
dQ_{+c}/dT_{+c}	First-order derivative of Q_{+c} with respect to T_{+c}
$dr(T_{+c})/dT_{+c}$	First-order derivative of $r(T_{+c})$ with respect to T_{+c}
$dU(T_{+c})/dT_{+c}$	First-order derivative of $U(T_{+c})$ with respect to T_{+c}
f(i,n)	$= n \cdot i/[1 - 1/(1 + i)^n]$
i	minimum rate of return (discount rate)
I	capital cost invested
K	concentration ratio defined as number of metric tons of material that must be processed to produce one metric ton of concentrate
M	mining cost per metric ton processed
M_o	mining cost per metric ton of ore
M_{o1}	value of M_o for process 1

Symbol	Description
M_{o2}	value of M_o for process 2
M_{stp}	current mining costs per metric ton delivered to low-grade stockpile
M_w	mining cost per metric ton of waste
n	number of years
NPV	net present value
NPV_o	net present value of previously scheduled production
NSR	net smelter return
$NSR(x_1, x_2)$	net smelter return, defined as returns from selling concentrate produced from one metric ton of ore with average grades x_1, x_2, less smelting charges
O	overhead cost per metric ton
O_o	overhead cost per metric ton of ore
O_{o1}	value of O_o for process 1
O_{o2}	value of O_o for process 2
O_{stp}	current overhead costs associated with mining and stockpiling one metric ton of low-grade material
O_w	overhead cost per metric ton of waste
p_1	proportion of metal 1 contained in concentrate that is paid for by smelter
p_2	proportion of metal 2 contained in concentrate that is paid for by smelter
P	processing cost per metric ton processed
P_o	processing cost per metric ton of ore
$P_o(T_{+c})$	processing cost per metric ton of ore processed, if plant capacity is T_{+c}
P_{o1}	value of P_o process 1
P_{o2}	value of P_o process 2
P_{stp}	current costs of stockpiling material that will be processed later
P_w	processing cost per metric ton of waste
Q_{+c}	quantity of metal contained in material above cut-off grade x_c: $Q_{+c} = T_{+c} \cdot x_{+c}$
$Q(x)$	quantity of metal in material for which the grade is greater than x

Symbol	Description
r	recovery, or proportion of valuable product recovered from the mined material
r_1	value of r for process 1
r_2	value of r for process 2
r_c	constant recovery after subtracting constant tail
r_{stp}	recovery expected at the time stockpiled material will be processed
$r(T_{+c})$	processing plant recovery, if plant capacity is T_{+c}
$r(x)$	process recovery for material of average grade x
R	refining, transportation, and other costs per unit of valuable material produced
R_1	value of R for process 1
R_2	value of R for process 2
R_{stp}	cost per unit of product sold
t	time, measured in years
T	tonnage to be mined from a new row of draw points
T_{+c}	tonnage above cut-off grade x_c
$T(x)$	tonnage of material for which the grade is greater than x
$U(T_{+c})$	utility of running the plant at T_{+c} capacity for one year
$U(x)$	utility of sending one metric ton of material of grade x to a given process: $U(x) = U_{dir}(x) + U_{opp}(x) + U_{oth}(x)$
$U_1(x)$	utility of sending one metric ton of material grade x to process 1
$U_2(x)$	utility of sending one metric ton of material grade x to process 2
$U_{dir}(x)$	direct utility (profit or loss) of processing one metric ton of material of grade x
U_{jk}	utility of mining block j in year k
$U_{jk,dir}$	direct utility of mining block j in year k
$U_{jk,opp}$	opportunity cost of mining block j in year k
$U_{jk,oth}$	other utility of mining block j in year k
$U_{opp}(x)$	opportunity cost or benefit of changing the processing schedule by adding one metric ton of grade x to the material flow
$U_{ore}(x)$	utility of mining and processing on metric ton of grade x
$U_{ore}(x_1, x_2)$	utility of sending one metric ton of material with metal grades x_1, x_2 to the processing plant

Symbol	Description
$U_{oth}(x)$	utility of other factors that must be taken into account in the calculation of cut-off grades
$U_{stp}(x)$	utility of stockpiling material of grade x
$U_{waste}(x)$	utility of mining and wasting one metric ton of material of grade x
V	value of one unit of valuable product
V_{stp}	dollar value of the product recovered from stockpile at the time product is sold
x	average grade
x_{+c}	average grade above cut-off grade x_c
x_{1e}	grade equivalent expressed in terms of metal 1
x_{2e}	grade equivalent expressed in terms of metal 2
x_c	cut-off grade
x_{c1}	cut-off grade 1, taking only operating costs into account
x_{c2}	cut-off grade 2, taking into account operating costs and undiscounted capital cost per metric ton
x_{c3}	cut-off grade 3, taking into account operating costs and discounted capital cost per metric ton
x_{c4}	cut-off grade 4, taking into account operating costs, discounted capital cost per metric ton and opportunity costs
x_s	selected cut-off grade
YCF	yearly cash flow

Index

NOTE: *f.* indicates figure; *n.* indicates (foot)note; *t.* indicates table.

Block or panel caving
- capital cost and cut-off grade, 59–60
- marginal cut-off grade and block design, 58–59
- marginal cut-off grade and draw point management, 58
- opportunity cost of increased size of block, 60–61
- and rate at which material is pulled, 57
- role of cut-off grades, 57
- and selectivity, 58
- and waste, 57

Breakeven cut-off grade, 23

Cut-off grade
- and costs and benefits, 1
- defined, 1
- and deposit models, 89–92, 90*f.*, 91*t.*
- and effect of increased mining capacity with fixed processing capacity, 71–74, 72*f.*
- and effect of increased processing capacity with fixed mining capacity, 75–77, 76*f.*
- and fixed costs, 63
- and fixed sales volume, 79–83, 81*f.*, 82*f.*
- fundamental equation, 93
- and grade-tonnage relationship, 6–7, 6*f.*
- and incremental capital expenditures, 66
- as iterative process, 93
- and leaching, 65
- lowering, and poor gold leaching results, 67–70, 68*f.*, 69*f.*
- and mine life, 1–2, 16–17
- minimum, 19–36
- and minimum return on investment, 64
- and net present value, 3
- and next step of processing, 1, 5
- and operating costs, 64
- opportunity cost of not using optimum, 33–36, 35*f.*
- optimization with opportunity costs, 14–15
- and overhead costs, 66
- and profitability, 1–2
- and reserves, 2
- review and revision of, 93
- and sunk costs, 64
- and sustaining capital, 65–66
- and variable costs, 63
- for various increases in mining or mill capacity, 85–87, 85*t.*, 86*t.*
- wide ramifications of, 93
- year of mining, 10, 11*f.*

Deposit models
- and cut-off grade, 89–92, 90*f.*, 91*t.*
- high-selectivity, 89, 90*f.*
- low-selectivity, 89, 90*f.*
- open-pit, 89–92, 90*f.*, 91*t.*
- underground, 89–92, 90*f.*, 91*t.*

Direct profit or loss, 6, 7
- base metal example, 8
- formulae, 7–8
- precious metal example, 8

The Economic Definition of Ore: Cut-Off Grades in Theory and Practice, 3

Fixed costs, 63

Fixed sales, 79
- with fixed mining rate and no processing constraint, 81–83, 82*f.*
- with fixed processing rate and no mining constraint, 80–81, 81*f.*
- with no mining or processing constraint, 79–80

Gold leaching
grade–tonnage curve, 67, 68*f.*
poor results from lowering cut-off grade, 67–70
relationship between leach recovery and solution ratio, 68–69, 69*f.*
Grade-tonnage relationship, 6
curves, 6–7, 6*f.*

Incremental capital expenditures, 66
Internal cut-off grade, 20

Lane, Kenneth F., 3
Leaching, and discounted recovery, 65. *See also* Gold leaching

Mill cut-off grade, 20, 25, 44
Mine cut-off grade, 20*n.*, 23
Mine life, 1–2, 16–17
Minimum metal content, 1
Minimum return on investment, 64
Mining capacity
cut-off grade for planned increase in, 86, 86*t.*
fixed, and effect on cut-off grade when processing capacity is increased, 75–77, 76*f.*
increasing, and effect on cut-off grade when processing capacity is fixed, 71–74, 72*f.*
optimizing, 73
planning changes in, 85–86, 85*t.*

Net present value (NPV), 3
and capacity constraints, 9
and capital cost, 59–60
optimization, 14, 15
relationship with opportunity cost and year of mining, 10, 11*f.*
Net smelter return (NSR)
copper–molybdenum example, 39–42
formulae, 38–39
mill or internal, 40
relationship to metal grades, 40, 41*f.*

Open pit mines
economic valuation of a pushback, 53–54
similarities in planning to underground mines, 55
Open pit mines, material at bottom, 22
base metal example, 23–24
breakeven cut-off grade, 23
mathematical formulation, 22–23
mine cut-off grade, 23
precious metal example, 23, 24*f.*
Operating costs, 64
Opportunity costs or benefits, 6, 9
and capacity constraints, 9–14
and constraints on mining or processing capacity (precious metals), 9–12, 11*f.*
and constraints on mining, milling, or refining capacity (base metals), 13–14
and constraints on smelter capacity or volume of sales (precious metals), 12
and cut-off grade optimization, 14–15
and not using optimum cut-off grade, 33–36, 35*f.*
and other costs, 15–17
relationship between NPV, opportunity cost, and year of mining, 10, 11*f.*
Optimizing processing plant operating conditions, 43
copper mine grinding circuit example, 45–51
grade–tonnage relationship for coming year, 45–46, 45*t.*, 46*f.*
mathematical formulation, 43–44
mill cut-off grade, 44
optimal plant capacity, 44
relationship between copper recovery and mill throughput, 48, 48*f.*
relationship between incremental utility and tonnage of mill feed, 48–51, 50*f.*, 51*t.*
relationship between operating cost per metric ton and tonnage processed per year, 47, 47*f.*

relationship between utility function and tonnage processed, 48, 49*t.*, 50*f.*
utility function, 43–44, 46, 48
Ore vs. waste, 19
base metal example, 22
internal cut-off grade, 20
mathematical formulation,19–20
mill cut-off grade, 20
mine cut-off grade, 20*n.*
precious metal example, 20–21, 21*f.*
See also Waste vs. low-grade stockpile
Overhead costs, 66

Polymetallic deposits, 37
cut-off values, 38
metal equivalent, 38, 40
metal equivalent (calculations), 40–42
mill or internal NSR, 40
net smelter return (copper–molybdenum example), 39–42
net smelter return (NSR), 38–39
relationship between NSR and metal grades, 40, 41*f.*
valuation formulae, 37–38
Processes, choosing between, 26
base metal example, 27–28
mathematical formulation, 26
precious metal example, 26, 27*f.*
Processing capacity
cut-off grade for increase in flotation plant capacity, 86*t.*, 87
cut-off grade for planned increase with fixed mill capacity, 86*t.*, 87
cut-off grade for planned increase with fixed mining capacity, 86*t.*, 87
fixed, and effect on cut-off grade when mining capacity is increased, 71–74, 72*f.*
increasing, and effect on cut-off grade when mining capacity is fixed, 75–77, 76*f.*
planning changes in, 85–86, 85*t.*
variance from idealized design balance with mining capacity, 71

Reserves, 2

Stakeholders, 2
balancing needs of, 16
Stockpiling, 2, 65. *See also* Waste vs. low-grade stockpile
Sunk costs, 64
Sustaining capital, 65–66
Symbols, 97–100

Underground mines
and blasting and haulage costs, 25
capacity constraints, 24
economic valuation of a stope, 54
mill cut-off grade, 25
minimum stope average grade, 24
similarities in planning to open pit mines, 55
and stope boundary material, 24–25
Utility, 5–6
defined, 5*n.*

Variable costs, 63
Variable recoveries
constant tail (base metal), 32–33, 33*f.*, 34*f.*
constant tail formulae, 32
formulae, 30
non-linear recovery (precious metal), 30, 31*f.*

Waste vs. low-grade stockpile, 28
formula, 28–30
See also Ore vs. waste

About the Author

Jean-Michel (J.M.) Rendu is an independent mining consultant with more than thirty-five years of experience in the mining industry. He is recognized worldwide as an expert in the estimation and public reporting of mineral resources and mineral reserves, and in geostatistics. In his former position as vice president of resources and mine planning with Newmont Mining Corporation for seventeen years, he was responsible for the management of all Newmont mining activities, including project reviews, staffing and support of corporate and mine-site mining groups, estimation and public reporting of mineral resources and mineral reserves, mine planning, and ore control.

J.M. has authored approximately fifty technical papers on deposit modeling, mine planning, methods and guidelines for estimation of resources and reserves, U.S. and international regulatory requirements for public reporting, and geostatistical theory and practice. He is also the author of *An Introduction to Geostatistical Methods of Mineral Evaluation*, a South African Institute of Mining and Metallurgy Monograph Series, first published in 1978.

He is an honorary professor at the University of Queensland, an adjunct associate professor at the Colorado School of Mines, an invited lecturer at École Polytechnique in Montreal, an outstanding instructor in mining engineering at the University of Wisconsin in Madison, and an external lecturer at the University of Witwatersrand in South Africa. J.M. has taught short courses on estimation of mineral resources and mineral reserves, public reporting of mineral resources and mineral reserves, cut-off grade calculation, and geostatistical theory and practice.

In addition to being a founding registered member of the Society for Mining, Metallurgy, and Exploration (SME), J.M. has chaired SME's Ethics Committee and the Resources and Reserves Committee, and was past director of the Mining and Exploration (M&E) Division. In addition, he is a founding member and U.S. representative of the Committee for Mineral Reserves International Reporting Standards. J.M. is an elected member of the U.S. National Academy of Engineering. He is a Fellow of both the Australasian Institute of Mining and Metallurgy and the South African Institute of Mining and Metallurgy.

J.M. was a recipient of the Henry Krumb Lecturer Award in 1992; the Presidential Award in 1992 and 2004; the Daniel C. Jackling Award in 1994; the M&E Division Distinguished Service Award in 2008; and the American Institute of Mining, Metallurgical, and Petroleum Engineers' Mineral Economics Award in 2008.